Isaac Stone

The elementary and complete examiner

Candidate's assistant

Isaac Stone

The elementary and complete examiner
Candidate's assistant

ISBN/EAN: 9783337281656

Printed in Europe, USA, Canada, Australia, Japan

Cover: Foto ©Andreas Hilbeck / pixelio.de

More available books at **www.hansebooks.com**

THE

ELEMENTARY

AND

COMPLETE EXAMINER;

OR

CANDIDATE'S ASSISTANT:

PREPARED TO '

AID TEACHERS IN SECURING CERTIFICATES FROM

BOARDS OF EXAMINERS,

AND

PUPILS IN PREPARING THEMSELVES FOR PROMOTION, TEACH-
ERS IN SELECTING REVIEW QUESTIONS IN NORMAL SCHOOLS,
INSTITUTES, AND IN ALL DRILL AND CLASS EXERCISES.

ISAAC STONE, A. M.,

PRINCIPAL KENOSHA HIGH SCHOOL.

A. S. BARNES & COMPANY,

NEW YORK AND CHICAGO.

1869

PREFACE.

CONSTANT labor of more than twenty-two years in the school-room has wrought some deep convictions in the mind of the Author. Many of these have been pleasant. But most of those pertaining to the *Examination of Teachers* have been painful,—painful from *sympathy* with the *Candidates ;* while deeply anxious to discharge his whole duty toward those who are to be the pupils of the successful Candidates, in commissioning none to teach except those who prove themselves well qualified for the high and responsible trust they are to assume.

The fact that *nine-tenths* of all the *teachers* he had examined failed in a greater or less degree, and that more than fifty per cent. failed sadly in some branches, led him, years ago, to conclude that he had misjudged their *ability,* or had placed the standard too high—exacting too much—as his Board of Education once hinted, where he was acting as City Superintendent. The late reports of County Superintendents, as found in the State Superintendent's report, *confirm* him, however, in his decisions, as they show that forty-nine fiftieths of all examined fail to secure *first grade certificates* in the *common branches.* A larger per cent. it is true, obtain the second grade. But the great majority accept third grade certificates. Their willingness thus to do ought to be a sufficient cause for rejecting them entirely ; justice to the children demands that this low grade should be banished from the land as an evil genius.

Should it appear to any that this work extends through too

many branches, and is too rigid, let such remember that each State, year by year, is demanding higher and still higher qualifications in all who are commissioned to labor in this noble profession.

By the advice of many *eminent educators*, the *Complete Examiner* is sent forth to the world on its errand of *benevolence*. It seeks no literary fame, claims no scientific merit. It simply begs to *aid* those who *need* its assistance. Receive it kindly, trustingly; it will tell no tales. Should it be the means of *aiding* PUPILS in preparing themselves thoroughly to be examined for promotion to a higher department, or CANDIDATES in procuring certificates for teaching, and thereby diminish the number of failures and lessen the degree of mortification therefrom, and as a consequence place a *higher order* of *talent* in the school-room, as disbursing agents of the infinite fountain of knowledge, the Author will feel amply rewarded for all his care in preparing this little work and sending it forth as a *love offering* to both teachers and pupils.

THE AUTHOR.

KENOSHA, March 10th, 1864.

CONTENTS.

	PAGE.
PREFACE,	3
CONTENTS,	5
HINTS TO CANDIDATES,	11

CHAPTER I.

ORTHOGRAPHY,	13
Elementary Sounds,	13
Consonants,	15
Syllables,	17
Rules for Spelling,	17
Punctuation,	18
Spelling,	19

CHAPTER II.

READING,	19
ACCENT,	21
EXPRESSION,	22

CHAPTER III.

GEOGRAPHY,	24
The Earth,	25
Land,	25
Water,	25
Maps,	26
Hemispheres,	26
Size and Motion of the Earth,	26
Latitude,	26
Longitude,	27
Zones,	27
Western Hemisphere,	28
Eastern Hemisphere,	28
Races of Men,	28
Stages of Society,	29
Government,	29
Political Divisions,	29
Religion,	29
Table of Races,	30
Table of the prevailing systems of Religion,	30
North America,	30
British Provinces,	31
United States,	31
Mexico and the West Indies,	33
Map of South America,	33
Europe,	34
Asia,	36
Africa,	37
Map of Oceanica,	38

PAGE.

CHAPTER IV.

Physical Geography,   39
    Continents and Islands,   40
    Mountains and Volcanos,   40
    Plateaus and Plains,   40
    Water,   40
    Ocean,   41
    Atmosphere and Moisture,   41
    Plants and Animals,   42

CHAPTER V.

Penmanship,   42

CHAPTER VI.

Book-Keeping,   44

CHAPTER VII.

Grammar,   46
    Language,   46
    Grammar,   47
    Words,   47
    Phrases,   48
    Sentences,   49
    Diagrams,   50
    General Rules,   50
    Classification of Sentences,   51
    Etymology,   52
    Of the Noun,   53
    Modification of Nouns,   54
    Person,   55
    Number,   55
    Cases,   55
    Pronouns,   56
    Adjectives,   57
    Verbs,   58
    Modes,   58
    Participles,   59
    Tenses,   59
    Conjugation of Verbs,   60
    Adverbs,   61
    Prepositions,   61
    Conjunctions,   62
    Exclamations,   62
    Words of Euphony,   62
    Syntax,   63
    Prosody,   65
    Grammatical and Rhetorical Signs,   65
    Versification,   66
    Figures,   66
    Abbreviations,   67

CHAPTER VIII.

Arithmetic,   69
    Notation,   69
    Numeration,   69
    Addition,   71

PAGE.

Subtraction, 71
Multiplication, 71
Division, 73
Longitude and Time, 74
Properties of Numbers, 75
Common Fractions, 77
Reduction of Fractions, 78
Addition of Fractions, 79
Subtraction of Fractions, 79
Multiplication of Fractions, 79
Division of Fractions, 80
Duodecimals, 80
Decimal Fractions, 81
Addition of Decimals, 81
Subtraction of Decimals, 82
Multiplication of Decimals, 82
Contractions in Multiplication, 82
Division of Decimal Fractions, 82
Contractions in Division, 83
Reduction of Common and Decimal Fractions, 83
Denominate Decimals, 83
Circulating Decimals, 83
Continued Fractions, 84
Ratio and Proportion, 84
Cause and Effect, 85
Compound Proportion, 85
Partnership, 86
Compound Partnership, 86
Per Centage, 86
Interest, 87
Compound Interest, 87
Discount, 88
Banking, 88
Bank Discount, 88
Commission, 89
Stocks and Brokerage, 89
Profit and Loss, 89
Insurance, 90
Endowments, 90
Annuity, 90
Assessing Taxes, 91
Custom House Business, 91
Equation of Payments, 91
Alligation, 92
Alligation Alternate, 92
Coins, Currency and Exchange, 92
Arbitration of Exchange, 93
General Average, 93
Tonnage of Vessels, 94
Involution, 94
Evolution, 94
Cube Root, 95
Arithmetical Progression, 96
Geometrical Progression, 96

		PAGE
Analysis,		97
Mensuration,		98
Gauging,		99
Mechanical Powers,		99
Pulley,		99
Uniform Motion,		100
Laws of Falling Bodies,		100
Specific Gravity,		100
Appendix,		101

CHAPTER IX.

History,		102
Modern History,		103
General History,		104
United States History,		104
American Independence,		105
Mexican War,		107
Secession and its Consequent Rebellion,		107

CHAPTER X.

Physiology,		109
Anatomy,		109
Structure of Man,		109
Chemistry of the Human Body,		109
The Bones,		110
Physiology of the Bones,		110
The Muscles,		111
Physiology of the Muscles,		111
The Teeth,		111
Digestive Organs,		111
Circulatory Organs,		112
Respiratory Organs,		112
Animal Heat,		113
Voice,		113
Skin,		113
Nervous System,		114

CHAPTER XI.

General Questions,		114

PART SECOND.

CHAPTER XII.

Botany,		128

CHAPTER XIII.

Algebra,		131
Signs,		131
Addition,		133
Subtraction,		133
Multiplication,		134
Division,		134
Factoring Polynomials,		135
Algebraic Fractions,		136
Equations of the First Degree,		138

PAGE.

Axioms, - 138
Solution of Equations, - 138
Elimination, - 139
Indeterminate Equations and Problems, 140
Inequalities, - 141
Powers and Roots, - 141
Extraction of the Square Root of Fractions, - 142
Extraction of the Square Root of Algebraic Quantities, - 142
Radical Quantities of the Second Degree, - 142
Equations of the Second Degree, - 143
Trinomial Equations, - 144
Permutations, Arrangement and Combinations, - 145
Binomial Theorem, - 145
Extraction of Roots, - 145
Arithmetical Progression, - 145
Geometrical Progression, - 146

CHAPTER XIV.

GEOMETRY, - 147
Definition of Terms, - 148
Explanation of Signs, - 148
Axioms, - 148
Theorems, - 149
Ratios and Proportions, - 149
Of the Circle, - 149
Book IV, - 150
Book V, - 150
Book VI, - 151
Book VII, - 151
Book VIII, - 152
Book IX, - 152

CHAPTER XV.

NATURAL PHILOSOPHY, - 153
Introduction, - 153
Preliminary Principles, - 153
Repellant Forces, - 153
Mechanical Principles, - 154
Principles of Gravitation, - 155
Principles of Molecular Action, - 157
Liquids, - 157
General Properties of Gases and Vapors, - 158
Air Pumps, - 160
Water Pumps, - 160
Buoyancy of the Atmosphere, - 161
Acoustics, - 161
Musical Sounds, - 162
Heat, - 163
Radiation of Heat, - 163
Reflection of Heat, - 163
Optics, - 166
Refraction, - 167
Optical Instruments, - 168
Microscopes, - 169
Structure of the Eye, - 169

PAGE.

Magnetism,   -   -   -   -   -   -   -   169
Directive Force of Magnets,   -   -   -   -   170
Electricity,   -   -   -   -   -   -   -   171
Dynamical Electricity,   •   -   -   -   172
Electro-Magnetism,   -   -   -   -   -   173

CHAPTER XVI.

CHEMISTRY,   -   -   -   -   -   -   -   173
    Chemical Nomenclature,   -   -   -   -   175
    Non-Metalic Elements,   -   -   -   -   -   176
    Hydrogen,   -   -   -   -   -   -   -   177
    Nitrogen,   -   -   -   -   -   -   -   179
    The Atmosphere,   -   -   -   -   -   -   -   179
    Chlorine,   -   -   -   -   -   -   -   180
    Iodine,   •   -   -   -   -   -   •   181
    Bromine,   -   -   -   -   -   -   -   181
    Fluorine,   -   -   -   -   -   -   -   181
    Sulphur,   -   -   -   -   -   -   -   182
    Phosphorus,   -   -   -   -   -   -   -   182
    Boron,   -   -   -   -   •   .   -   183
    Silicon,   -   -   -   -   -   -   -   183
    Carbon,   -   -   -   -   -   -   -   183
    Combustion,   -   -   -   -   -   -   184
    Metalic Elements,   -   -   -   -   -   185
    Sodium,   -   -   -   -   -   -   •   186
    Ammonium,   -   -   -   -   -   -   186
    Barium and Strontium,   -   -   -   -   187
    Calcium,   -   -   -   -   -   -   187
    Magnesium,   -   -   -   -   :   -   -   187
    Alluminium,   -   -   -   -   -   •   187
    Glass and Pottery,   -   -   -   -   -   187
    Iron,   -   -   -   -   -   -   -   188
    Manganese and Chromium,   -   -   -   -   -   188
    Cobalt and Nickel,   -   -   -   -   -   188
    Zinc and Cadmium,   -   -   -   •   -   -   188
    Lead and Tin,   -   -   -   -   -   -   188
    Copper and Bismuth,   -   -   -   -   189
    Antimony and Arsenic,   ⌣   -   -   .   -   189
    Mercury,   -   -   •   ⌣   -   ⸜?   -   190
    Silver,   -   -   -   -   -   -   -   190
    Gold,   •.   -   -   -   -   -   -   190
    Platinum,   -   -   -   -   -   -   191
    Organic Chemistry,   -   -   -   -   -   -   191

CHAPTER XVII.

SCIENCE OF GOVERNMENT,   -   -   •   -   -   193

CHAPTER XVIII.

MUSIC,   -   -   -   -   -   -   •   -   197

CHAPTER XIX.

THEORY AND PRACTICE,   -   -   -   -   •   200

CHAPTER XX.

MISCELLANEOUS QUESTIONS,   -   -   -   -   202

HINTS TO CANDIDATES.

The law demands that you should pass a satisfactory *examination* before you can be commissioned to *teach*.

Justice and your own conscience coincide with the demands of the law.

You should never willingly engage to teach a branch in which you are deficient. Never ask an *examiner* to license you to teach a branch which you have never thoroughly mastered.

With the *aid* of the *Complete Examiner* you have the means before you of determining *when* and in *what* you will be examined.

By being thorough and rigid in questioning yourself by the use of *this work*, you may not only save the examiner the painful emotions in rejecting you, but also your own mortification from the disgrace of your failure.

Observe the *remarks* at the head of each chapter. Be especially thorough in the *first* few chapters.

There may be many questions that you can *answer mentally*, as fast as you can read them.

It would be safer and more beneficial to you to write out the answers to the balance of the questions, in your own words, if you choose.

Answer every question, if possible, without turning to any reference in the *text-books*. Be as *calm*, as *self-possessed*, and as *much* at *ease*, during your *public examination*, as in *your own room*. You may be able, in many places, to pass a satisfactory examination without a complete knowledge of all the branches touched upon in this work. But when you can give positively correct answers to all the questions in the COMPLETE EXAMINER, you need not hesitate to present **yourself** for examination before *any Board of Examiners.*

THE COMPLETE EXAMINER.

ORTHOGRAPHY.

CHAPTER I.

The following questions on Orthography are taken from Worcester's Un-abridged Dictionary, as the most complete source of knowledge on the subject. Yet a good knowledge of it may be obtained from several Spellers and Grammars.

NOTE.—Make the sounds which the characters represent in the following questions wherever it is more convenient than to write out the description.

1. Define Orthography. (See Worcester's Un. Dic.)
2. What is a Letter?
3. What is an Elementary Sound? Syllable?
4. What is a Word? Sentence? Paragraph? Chapter?

ELEMENTARY SOUNDS.

1. How many Elementary Sounds in our Language?
2. How many Letters have we to represent the sounds?
3. Into how many Classes are these elementary sounds divided?
4. What is a Vocal, Sub-vocal, and Aspirate?
5. How many Vowels have we?
6. How many Semi-vowels? How many Aspirates?
7. How many Sounds do the vowels represent?
8. How many Sounds do the semi-vowels represent?
9. What is a Diphthong?
10. What is a Triphthong?
11. What sound has the letter a in fate, and ai in pain?
12. What sound has a in fat, man? Far, calm?
13. What sound has a in fare, pair?
14. What sound has a in *fast, grass*? *Fall, walk*?
15. What sound has a in *liar*? *Palace, cabbage*?

16. What sound has *e* in *mete, seal*? *Met, sell*?

17. What sound has *e* in *there, heir*? *Her, fern*?

18. What sound has *e* in *brier, college*?

19. How does *e* in *college* differ from *a* in *cabbage*?

20. What is the sound of *i* in *pine, mild*? *Miss, pin*?

21. What is the sound of *i* in *police, marine*? *Sir, virtue*?

22. What is the sound of *i* in *elixir, ability*?

23. What sound has *o* in *sore, ton*? *Odd, dove*?

24. What sound has *o* in *prove, soon, nor, form*?

25. What sound has *o* in *come, wrong, actor, purpose*?

26. What sound has *u* in *pure, tube*? *Hut, hurry*?

27. What is the sound of *u* in *bull, push*? *Fur, hut*?

28. What sound has *u* in *true, rude*? *Sulphur, deputy*?

29. What is the sound of *y* in *type, symbol*? *Myrrh, truly, envy*?

30. When are *w* and *y consonants*? When *vowels*?

31. Give a set of words that contain the different elementary sounds in the language.

32. In the following words, which letters are *Vowels*, which *Semi-vowels*, and which *Aspirates*:—*And, great, made, fame, sad, mate, life, six, vice, zebra, sup, bid, bag, pare, when, this, shall, ocean, link*?

33. In the above examples which vowels are *long*? Which *short*?

34. In the word *union* are the *i* and *u vowels* or *consonants*?

35. In the word *one* is *o* a *vowel* or *consonant*?

36. In the word *righteous* is the *e* a *vowel* or *consonant*?

37. How *many letters* have we then that are *always consonants*?

38. What is a *Digraph*?

39. What is an *Improper Triphthong*?

40. In the words *boil* and *boat* which has a *Digraph*? Which a Diphthong?

41. In the words *beauty* and *buoy*, which has the Proper Triphthong? Which the Improper Triphthong?

42. What sound has *ew* in *mew*? *oi* in *boil*?

CONSONANTS.

1. What is a Consonant?

2. How many sounds has *c*? Give examples.

3. How many sounds has *ch*? Give examples.

4. What is the sound of *g* in *get, give, giant*?

5. How many sounds has *s, x*?

6. What sound has *th* in *this, then, think, pith*?

7. How is *tion* pronounced in *notion*, and *sion* in *pension*?

8. How is *sion* pronounced in *vision*, and *cean* and *cian* in *ocean* and *logician*?

9. Give examples and tell the different sounds of *cial, sial,* and *tial*.

10. What is the sound of *ceous, cious,* and *tious*? Give examples.

11. What is the sound of *geous* and *gious*? Give examples.

12. What sound has *qu* in *queen*?

13. In the word *when*, which letter is sounded first, *w* or *h*?

14. What is the sound of *ph* in *phantom*?

15. How many sounds have the following Diphthongs:—*ea, eu, ew, ia, ie, io, oi, ou, ow, oy, ua, ue, ui, uo*?

16. Give examples and tell what sound each of the following Diphthongs has:—*ae, ai, ao, au, aw, ay; ea, ee, ei, eo, ey; ie, oa, oe, oo, ow*.

17. In the last two questions what Diphthongs are Proper, and what Improper?

18. Into how many classes are the Consonants divided?

19. What is a Mute?. Which of the consonants are Mutes?

20. Name the Semi-vowels.

21. What is a Liquid? Why? Name the Liquids.

22. Which of the consonants are Dentals? Why so called?

23. Which letters of the Alphabet are Palatals?

24. Which are Labials? Which are Nasals?

25. Which letters are called Gutturals? Why?

26. What sound has *b* preceded by *m* in the same syllable?

27. What sound has *c* when it comes after the accent and is followed by *ea, ia, io,* or *eous*? Give examples.

28. What sound has *d* in *healed* and *mixed*?

29. What sound has *p* in *up*?   In all other cases?

30. What sound has *g* followed by *r* at the beginning of a word?

31. How many *g* sounds in the word *longer* (the comparative of *long*,) and *longer* (one who longs)?

32. What sound has *gh* at the beginning of words?

33. What sound usually at the end of words?   Give examples.

34. What sound have they in the words *plough, nigh, laugh*?

35. The combination of letters *ough* is said to have seven sounds; what are they?   Give examples.

36. What sound has *ght* terminating a word?

37. What sound has *h* in *her, heir, honor*?

38. What sound has *k*?   When is it silent?

39. How many sounds has *l*?   Is it ever silent?

40. Is *m* ever silent?   Give examples.

41. What sound has *n* in *man, no, angle, thanks*?

42. Give examples when *n* is silent.

43. What sound has *p* in *pit, psalm*?

44. How many sounds has *ph*?   Give examples.

45. How many sounds has *q*?   Give examples.

46. How many sounds has *r*?   Is it ever silent?   What effect does it have on the short sound of the vowels?   Give examples.

47. How many sounds has *s*?   Give examples.

48. What sound has *s* at the beginning of a word?

49. What sound has *s* usually at the end of a word?

50. When has *s* the sound of *sh*?   When *zh*?

51. How many sounds has *t*?   What sound in *notion*, militia?

52. What is the sound of *th* in *truth, truths, bath, baths*?

53. Is *v* ever silent?   Give examples.

54. When is *w* a consonant?   Is it ever silent?   Give examples.

55. How many sounds has *x*?   What sound in *exalt*?

56. What sound has *x* in *luxury, fluxion*?

57. How many sounds has *y*?   Is it ever silent?.

58. How many sounds has *z*?   Is it ever silent?

59. How many sounds has *j*?   Is it ever silent?

60. What sound has *z* in *glazier, azure*?

61. What sound has *u* in *nature, educate*?

SYLLABLES.

1. How many syllables are there in every word? Ans.—Usually as many as there are distinct sounds made in pronouncing it.

2. What is a word of one syllable called? A word of two syllables? Of three syllables? Of four or more syllables? Give examples of each.

3. What is the rule for dividing a word at the end of a line? Ans.—Never divide a syllable.

4. What is a Simple Word? A Compound Word? A Primitive Word? A Derivative Word? Give examples.

5. Should the Hyphen always be used in forming Compound Words? If not, when and why?

6. What is a Prefix? A Suffix?

7. How are Derivative Words formed? Ans.—By correctly uniting Prefixes and Suffixes.

8. What is Spelling, and how would you teach it?

RULES FOR SPELLING.

1. What is the rule for *e* final?

Ans.—*E* final is dropped before the addition of a suffix beginning with a vowel; as, *recite, recital; blame, blamable.*

Exception.—Words ending in *ge* or *ce*, usually retain *e* final, before *able* and *ous;* as, *peace, peaceable.*

2. What is the rule for *e* final before *ly, less, ness, ful,* and generally *ment?*

Ans.—The *e* must be retained; as, *nice, nicely; hope, hopeful.*

Exception.—*Awe, due, true, judge,* and five more words. Which are they?

3. What is the rule for *y* final, upon the addition of a suffix?

Ans.—The *y* is commonly changed into *i*, except before *ing;* as, *mercy, merciful; holy, holiness.*

Exception 1.—*Y* is sometimes changed into *e;* as, *duty, duteous.*

Exception 2.—When *y* is preceded by a vowel in the same syllable, except in *lay, say,* and *pay*, it remains unchanged; as, *boy, boyish.*

4. What is the rule for verbs ending in *ie*, before the suffix *ing?*

Ans.—The *ie* is changed to *y;* as, *tie, tying; vie, vying.*

5. What is the rule for forming derivatives from monosyllables and words accented on the last syllable, ending with a single con-

sonant, or sound of a single consonant preceded by a single vowel?

Ans.—Such words double the last consonant before an additional sylla-ble, beginning with a vowel; as, *pet, petted.*

6. There are a few exceptions to this rule, in regard to Di-graphs; what are they?

7. What is the rule for *t* or *s* preceding *e* final -in such words as admit *ion?*

Ans.—The *e* is dropped and *ion* is added.

8. Give the rule for derivatives from words ending in *ate.*

Ans.—They drop *te* and take *ble* or *cy;* as, *estimate, estimable.*

9. What is the rule for *y* before the suffixes, *ous, al,* and *able.*

Ans.—The *y* is commonly changed to *i* and usually retained; as, *rely, reliable.*

10. What is the rule for derivatives from words ending in *fy?*

Ans.—They change the *y* into *i* and take *cation.*

Exception.—A few words ending in *fy*, drop *y* and take *action;* as, *pet-rify, petrifaction.*

11. What is the rule for words ending in *ize?*

Ans.—They drop *e* and take *ation;* as, *civilize, civilization.* Also a few words not ending in *ize,* take *ation;* as, *sense, sensation.*

12 Form all the derivatives you can from the word *press,* and give rules for their formation, and define each derivative.

13. Analyze the following words by giving the Elementary Sounds:—*Mete, balm, rough, high, thought, laugh, beauty, phthisic, heir, bouquet, old, young.*

14 In the above words, which letters are vocals, which sub-vocals, and which aspirates?

PUNCTUATION.

1. What important rule or rules can you give for the use of Capital Letters? A hint:—The first word of every *entire sen-tence; titles* of *honor* and *respect;* every *proper name;* every *ap-pellation* of the *Deity;* the *first word* of every line in *poetry;* the words *I* and *O;* the principal words in the *titles* of *books;* the *first word* of a *direct quotation* (when the quotation forms a complete sentence by itself) *should all begin* with a *capital;* and every important word *may* begin with a capital.

SPELLING.

Write the following list of words and make all the corrections necessary, and tell which represent animals that are *wild, gregarious, amphibious, ferocious, carnivorous, graminivorous, omnivorous, predatory, ruminating, migratory, venomous,* and *hybernating.*

Name the countries in which each is native; which are valuable for *food;* which for their *fur;* which for their *skins;* which for their *ivory;* which for oil; which are serviceable to man, and in what way:—

Ape, antilope, Babboon, bare, beaver, Buffalo, bizon, caff, cammell, Shamme, catt, coult, koogar, cow, dear, dogg, Elephant, ermin, fox, gazel, gote, horse, hieny, ibex, iknuman, jackkall, kitten, leppard, marten, lyon, munkey, muskrat, ottar, ox, panthar, pecary, rabit, seel, sheep, scunk, tigar, weesel, whale, wolf, zebra.

(The above words by permission, are taken from "Sherwood's Practical Speller and Definer," page six,—the best work of the kind ever published. I have designedly changed the Orthography of most of the words taken. But the *excellent plan* of *defining* is Sherwood's.)

Note—Every one who has had any experience in examining teachers or pupils, has found the candidates wofully deficient in Orthography. It is the more culpable from the fact, that any candidate for promotion to a Grammar Department, a High School, or for a Teacher's Certificate, can prepare himself or herself in this most important and *fundamental branch,* in a few hours. Hence all who neglect a thorough knowledge of the branch should be rejected.

CHAPTER II.

READING.

SUGGESTION.—Reading is a branch in which nineteen out of every twenty are deficient. Yet this is an *age* of reading, emphatically so. Notwithstanding this, we hear the testimony from every side, that "*there are but few good readers.*" How important then that we should have clear and distinct principles in Elocution, and that every person who is a candidate for a Teacher's License should be a complete master of every principle, and be able to impart the instruction in this branch to his pupils without stint.

N. B.—The following references in the questions on Reading are to "*The National Fifth Reader,*" by Parker & Watson.

1. What is Reading? What is Elocution? (Page 15.)

2. What does it embrace? What are the characteristics of good Elocution? (Page 15.)

3. What is Orthoepy?   What does it embrace?   (Page 15.)

4. What is Articulation?   (Page 15.)

5. What are the Oral Elements?   (Page 15.)

6. How are the Oral Elements produced?   (Page 15.)

7. What are the principal organs of speech?   (Page 16.)

8. What is Voice and how is it produced?   (Page 16.)

9. What directions should be given to pupils in regard to the position of their bodies while reading?   (Page 17.)

10. How would you teach the pupil the *oral elements ;* would you require him to pronounce the word, or to utter each element by itself and then pronounce the word distinctly?   (Page 17.)

11. What are Cognates?   How would you teach them?   (Page 18.)

12. What advantage is there in teaching the pupils to spell by sounds?   (Page 20.)

13. Name the errors which are usually heard in Articulation?   (Pages 21 and 22.)

14. Why is the following difficult of articulation :—

1. "He *accepts* the office, and *attempts* by his *acts* to conceal his *faults.*"

2. "He was attacked with spasms and died miserably by the road-side."

3. "For the hundredth time, he spoke of lengths, breadths, widths, and depths."

4. "Theodore Thickthong thrust three thousand thistles through the thick of his thumb."   (Pages 22 and 23.)

5. "He said, ceaseth, approacheth, and rejoiceth."   (Page 24.)

15. What is Syllabication?   What is a Syllable?   (Page 25.)

16. What is a Monosyllable?   A Dissyllable?   (Page 25.)

17. What is a Trisyllable?   A Polysyllable?   (Page 25.)

18. Which is the Ultimate syllable?   The Penult?   (Page 25.)

19. Which is the Antepenultimate?   The Pre-antepenultimate?   (Page 25.)

20. Give examples for each of the last five questions.

21. What is the rule for pronouncing words that commence with consonants?   (Page 26.)

22. What is the rule for pronouncing words that end with consonants?   (Page 26.)

23. What is the rule for pronouncing when one word ends and

the next begins with the same consonant; as, "*it will pain nobody*"? (Page 26.)

24. What is the rule for the utterance of the final elements:— *b, p, d, t, g* and *k;* as, "I took down my hat-t and put it on my head-d." (Page 27.)

25. What is the rule for pronouncing the unaccented syllables? (Page 27.)

NOTE.—Illustrate the above rules with suitable examples.

ACCENT.

26. What is Accent? What is the general rule for Accent?

ANS.—All the words of our language of two or more syllables have one syllable accented, and most polysyllables have both a Primary and a Secondary Accent.

27. In dissyllabic nouns where is the accent placed?

ANS.—On the Penult.

28. In dissyllabic verbs, where is the accent?

ANS.—On the last syllable.

29. In words ending in *sion* and *tion*, which syllable takes the accent?

ANS.—The Penultimate; as, *dissen'sion.*

30. In words ending in *ia, iac, ial, ian, eous,* and *rous,* which syllable takes the accent?

ANS.—Commonly the preceding; as, *regal'ia, imper'ial.*

31. Words ending in *acal* and *ical* have the accent on what syllable?

ANS.—Antepenultimate; as, *poet'ical.*

32. On what syllable do words ending in *ic* have the accent generally?

ANS.—On the Penultimate; as, *algebra'ic.*

33. Words of three or more syllables, ending in *ear, cal, tude, efy, ety, ity, graphy, logy, ulous, inous, erous, owrous, ative,* &c., have the accent on what syllable generally?

ANS.—On the Antepenultimate.

34. In the same sentence or adjoining one where there is a reference of one word to another, with perhaps a change in the prefix, is there a change of accent?

ANS.—There is; as, to *give* and *forgive; probability* and *plausibility.*

35. As authorities for the above rules for *accent,*—see Worcester's and Webster's Dictionaries, on *"Accent."*

36. What marks are used to show on what syllables the Primary and Secondary accents fall? (Page 29.)

37. Tell on what syllables the primary and secondary accents fall, in the following :—The impenetrability and indestructability are two essential properties of matter." (Page 29.)

38. What is the rule for accent on *nouns, adjectives,* and *verbs?* (Page 30.)

39. Mark the accented syllables in the following words and give the parts of speech :—

"Why does your absent friend absent himself?" "Did he abstract an abstract?"

Note the mark of accent, and accent the right syllable.

Buy some cement and cement the glass. Desert us not in the desert. If they rebel and overthrow the government even the rebels themselves can not justify the overthrow. In August, the august writer entered into a compact to prepare a compact discourse. (Page 30.)

40. What is the rule for contrast? (Page 30.)

41. Note the accent in the following sentences :—

"He must increase, but I must decrease. This corruption must put on incorruption ; and this mortal must put on immortality." (Page 30.)

EXPRESSION.

42. What is Expression?

ANS.—It is the *soul* of elocution.

43. What does it embrace?

44. What is Emphasis? Inflection? Slur? (Page 31.)

45. What is Modulation? Monotone? Personation? Pauses?

46. Give four rules for the use of Emphasis. (Page 32.)
    Give examples to illustrate each rule. (Page 32.)

47. What rules apply for the use of Slur? (Page 35.)
    Give any examples. (Page 36.)

48. How many Inflections are there ? Name them. (Page 39.)

49. How are these inflections indicated in the books? (Page 39.)

50. What is the *rising* inflection, and the *falling?* (Page 39.)

51. What is the Circumflex? Give examples for each. (Page 39.)

52. What is meant by the *slide* of the *voice?* (Page 40.)

53. How many parts does the slide consist of, and how many things are necessary to the perfect formation of the slide? (Page 40.)

54. Give the rule for the rising inflection, and falling inflection, and an example for each. (Page 41.)

55. Is there any exception to the above rules? Give it. (Page 41.)

56. Give the rule for the inflections in questions and clauses connected by the disjunctive, *or.* (Page 42.)

57. Give the rule for inflections, when words or clauses are contrasted. Illustrate by examples. (Page 43.)

58. What inflection does the language of *concession, politeness, admiration, entreaty,* and *tender emotions,* usually require?

Ans.—The *rising.* (Page 44.)

59. What inflection has the language of *command, rebuke, contempt, exclamation,* and *terror?*

Ans.—Falling.

60. What inflection has a succession of particulars?

61. What inflection does *emphatic* repetition and the pointed enumeration of particulars require?

Ans.—Falling. (Page 46.)

62. How is the language of irony, sarcasm, derision, condition, and contrast, marked?

Ans.—By the circumflex. Give examples. (Page 46.)

63. What is Modulation? What does it embrace? (Page 47.)

64. Define Pitch. (Page 47.) How many general distinctions of Pitch?

Ans.—High, Low, and Moderate.

65. Define High Pitch. Low Pitch. (Pages 47 and 48.)

66. Illustrate Moderate Pitch, and Low Pitch, by examples. (Page 49.)

67. Define Force. How many *degrees* of force? (Page 50.)

68. Define Loud Force, Moderate Force, and Gentle Force. (Pages 51 and 52.)

69. Define Quality. How many kinds of tone are used in reading and speaking? (Page 52.)

70. What is Pure Tone? Illustrate by an example. (Page 52.)

71. Define Orotund, and illustrate by an example. (Page 53.)

72. Define Aspirated Tone, and give an example. (Page 54.)

73. Define Guttural by an example. (Page 54.)

74. Define the Tremulous or Tremor.

75. Define Rate. How many *degrees?* (Page 56.)

76. Define Quick Rate, Moderate Rate, and Slow Rate, and give examples to illustrate.

77. Define Monotone. Give examples. (Page 59.)

78. Define Personation. (Page 60.)

79. What are Pauses in elocution? (Page 61.)

80. What are the general rules for the use of Pauses? Also the rule for Suspensive Quantity. (Pages 61, 62 and 64.)

81. Read the following sentence, so that it will make a *temperance* speech, and an *anti-temperance speech :—*

"The person who is in the daily use of intoxicating liquors, if he does not become a drunkard, will be in danger of losing his health and character."

CHAPTER III.

GEOGRAPHY.

The following references in the questions on Geography are to CAMP'S HIGHER GEOGRAPHY,—a *work of rare merit* "arranged to accompany Mitchell's Series of Outline Maps," but can be used independently, as the book is complete in itself. (P. stands for page.) (C. for column.)

Definitions of Mathematical terms used in Geography.

1. What is a Sphere? What is the Diameter of a sphere? P. 7, C. 1.

2. What is the Circumference of a sphere? What is the Axis of a sphere? P. 7, C. 1.

3. What are the Poles? What is a Circle? P. 7, C. 1.

4. What are the *great* circles of a sphere? What is an Arc? P. 7, C. 2.

5. How are *arcs* measured? P. 7, C. 2.

Geographical Definitions.

THE EARTH.

1. What is the Earth?

ANS.—A Planet.

2. What is a *planet?*

ANS.—A body revolving around the sun.

3. What are some of the planets?

ANS.—Mercury, Venus, Mars, Jupiter, Saturn, Uranus, Neptune, and a large number of others, called *asteroids.*

4. What is Geography, and what is the *origin* of the word? P. 8.

5. What is the form of the Earth? The proofs?

6. Of what does the surface of the earth consist? P. 8.

LAND.

1. What portion of the earth is land? What portion water? P. 8.

2. What are the principal divisions of land? P. 8.

3. What is a Continent? Island? Peninsular? P. 8.

4. What is an Isthmus? Cape? Promontory? Mountain? P. 8.

5. What is a Volcano? Hill? Plain? Valley? Desert? Oasis? P. 8.

6. What is a Shore or Coast? Plateau? P. 8.

WATER.

1. How is the water divided? P. 9.

2. What is an Ocean? How many Oceans are there? P. 9.

3. What is a Sea? Archipelago? Gulf or •Bay? Strait? Channel? Sound? Lake? River? Creek? P. 9.

4. Which is the right bank of a River? Which the left? P. 9, C. 2.

5. How are rivers formed? What is the source of a river? P. 10.

6. What is the mouth of a river? What is the bed of a river? P. 10.

MAPS.

1. Define a map. What does a map represent? P. 10.

2. In what direction is the top of the map supposed to be? P. 10.

3. What direction is the bottom of the map? The right hand? The left hand? What are these directions called? P. 10.

HEMISPHERES.

1. Define Hemisphere? How many hemispheres are there? P. 10.

2. Which is the Eastern Hemisphere, and what does it represent? P. 11.

3. What is the Western, and what does it represent? P. 11.

4. Which Hemisphere contains most land? P. 11.

5. Which most water? P. 11.

6. For what is the Eastern Continent distinguished? P. 11.

7. For what is the Western Continent noted? P. 11.

SIZE AND MOTIONS OF THE EARTH, EQUATOR AND CIRCLES.

1. What is the size of the Earth? What is the axis of the Earth? P. 11.

2. What motions has the Earth? P. 11.

3. What is the Equator? How does it divide the Earth? P. 11.

4. What are the Tropics, and why are they $23°\frac{1}{2}$ from the Equator? P. 11.

5. What are the Polar Circles? Why are they $23°\frac{1}{2}$ from the Poles? P. 11.

LATITUDE AND LONGITUDE.

1. What is Longitude? How is it represented on the map? P. 12.

2. What are these lines called? What do the figures attached to them show? P. 12.

3. Where do we begin to reckon the degrees of latitude? P. 12.

4. How many degrees between the Equator and each Pole? P. 12.

5. What is said of places either North or South of the Equator? P. 12.

6. What of places on the Equator? P. 12.

7. What is the length of each degree of latitude? P. 12.

8. How then may we learn the distance of a place from the Equator? P. 12.

LONGITUDE.

1. What is Longitude? What are Meridians? P. 12.

2. From what meridian do we usually reckon Longitude? P. 12.

3. What is the custom of different nations in this respect? P. 12.

4. Where are the degrees of longitude usually marked on the map? P. 12.

5. How many degrees of longitude are there? P. 12.

6. How many degrees then around the Earth? P. 12.

7. What longitude have places on the first Meridian? P. 12.

8. How can you tell whether the longitude of a place be East or West? P. 12.

9. What is the length of a degree of longitude? P. 12.

10. Give the *table* of *longitude*, showing the number of miles in a degree of *longitude* on a parallel of latitude, for every five degrees, from the Equator to the Poles,—sixty geographical miles being taken equal to sixty-nine and a quarter statute miles? P. 23.

ZONES.

1. What are zones, and what does the word mean? How many zones are there? P. 13.

2. What is the North Frigid? What the South Frigid? P. 13.

3. What is the climate of the zones? What the productions? P. 13.

4. What animals are found? What can you say of the inhabitants? P. 13.

5. What does the North Temperate Zone embrace? South Temperate? P. 13.

6. What is the climate of the Temperate Zones?

7. What minerals are found? P. 13.

8. For what is the North Temperate Zone more particularly distinguished?

9. Where is the Torrid Zone situated? What is its climate? P. 13.

10. What are the productions of this zone? What animals? P. 13.

11. Describe its inhabitants? To what is the Torrid Zone subject? P. 13.

WESTERN HEMISPHERE.

1. Give the seas, gulfs and bays on the map of the Western Hemisphere.

2. Mention all the straits and islands on the same map. P. 19.

3. Give also the Peninsulars and Capes. P. 19.

4. Give the Mountain ranges and their direction. P. 19.

5. What five large lakes in North America are connected, and discharge their waters into the Gulf of St. Lawrence? P. 19.

6. What are the principal rivers of North America? P. 19.

7. Name the principal rivers of South America? P. 19.

EASTERN HEMISPHERE.

8. Name the Seas, Bays and Gulfs on the map of the Eastern Hemisphere. P. 19.

9. Give the names and direction of the principal channels and straits. P. 20.

10. Locate the principal Islands, Capes and Peninsulars. P. 20.

11. What Mountains, Lakes and Rivers? P. 20.

RACES OF MEN.

1. How are mankind divided? P. 21.

2. How is the European Race distinguished? P. 21.

3. What Nations are included in the European Race? P. 21.

4. How is the Asiatic race distinguished? P. 21

5. What nations does it include? P. 21.

6. How is the American Indian race distinguished? P. 21.

7. What nations does it include? P. 21.
8. How is the Malay race distinguished? P. 21.
9. What nations are included in this race? P. 21.
10. How is the African race distinguished? P. 21.
11. What nations does it include? P. 21.

STAGES OF SOCIETY.

1. On what does the social condition of men depend? P. 21.
2. What do the different degrees of advancement among men in these particulars form? How many of these are there? P. 21.
3. What can you say of savage nations? P. 22.
4. What is the condition of half-civilized nations? P. 22.
5. What nations are civilized? Give examples. P. 22.
6. For what are enlightened nations noted? P. 22.
7. What nations are enlightened? How distinguished? P. 22.

GOVERNMENT.

1. What are the different forms of government? P. 22.
2. What is a Monarchy? How many kinds? P. 22.
3. What is an absolute monarchy? Limited monarchy? P. 22.
4. What is an Aristocracy? Democracy? P. 22.

POLITICAL DIVISIONS.

1. What are the Political divisions of the Earth? P. 22.
2. What is an Empire? Republic? Kingdom? P. 22.
3. What is the Chief Officer of a Republic called? P. 22.
4. How is he elected?
5. How are Empires, Kingdoms, and Republics subdivided? P. 22.
6. How are States subdivided?

RELIGION.

1. What are the principal systems of Religion? P. 22.
2. What Nations are called Christians? P. 23.
3. How are christians subdivided? P. 23.
4. Who are Mohammedans? Jews? Pagans? P. 23.

TABLE OF RACES.

1. How many souls do the Caucasian race number? P. 23.
2. How many do the Asiatic or Mongolian? P. 23.
3. How many do the African or Negro? P. 23.
4. How many do the Malay? P. 23.
5. How many do the American or Indian? P. 23.

TABLE OF THE PREVAILING SYSTEMS OF RELIGION.

1. How many do the Jews number? P. 23.
2. How many do the Christians number? P. 23.
3. How many do the Pagans and Mohammedans number? P. 23.

NORTH AMERICA.

1. What part of the globe does North America comprise? P. 27.
2. What is its length? Breadth? P. 27.
3. What Ocean bounds North America on the north? P. 27.
   On the East? West? What isthmus connects it with South America? See map North America.
4. Bound the different divisions of North America.
5. Draw a map of North America.
6. What sea between North and South America? P. 27.
7. Name the principal seas, Gulfs, and Bays of N. A. P. 27.
8. What strait between Asia and N. A.? North of Brit. A.? P. 28.
9. What strait between B. A. and Greenland? P. 28.
10. Name the principal straits and sound, and locate them. P. 28.
11. Locate the principal islands, and name them. P. 28.
12. Name and locate the principal *capes* and *peninsulas*. P. 28.
13. Name and give the directions of the Mountain chains. P. 28.
14. Name and locate the lakes. Rivers.
15. How would you go by boat and car from Chicago to San Francisco?
16. How would you go by steamer from Chicago to Liverpool?

17. How does North America rank in size among the other divisions? P. 31.

18. What division is most mountainous? P. 31.

19. What division is most level? P. 31.

20. Who inhabit Russian America? What is its capital? P. 31.

21. Are there any Volcanoes in it? Name them. P. 31.

BRITISH PROVINCES.

1. Bound British America. See map No. 2.

2. In what part is Hudson's Bay Territory? Labrador?

3. Bound Upper Canada. Lower Canada. P. 34.

4. Bound New Brunswick. Nova Scotia. Newfoundland. P. 34.

5. Name the Oceans, Seas and Bays. P. 34.

6. Name the Straits and Channels, Islands, Capes and Lakes. P. 34.

7. What rivers flow into James Bay? P. 34.

8. What river is the boundary between New Brunswick? P. 34.

9. What river connects Lake Erie and Ontario?

10. Name the chief rivers and give their directions.

11. Draw a map of the British Provinces.

UNITED STATES.

1. What country bound the U. S. on the North?

2. What ocean on the East? What Gulf and country on the South?

3. What ocean on the West? What is the latitude of the U. S.?

4. What is the longitude? How many states are there?

5. How many territories are there? Name the States and Territories.

6. What states border on the Atlantic? On the Gulf of Mexico?

7. On the Pacific? On the Great Lakes? What States lie west of the Mississippi? What States are separated by the Connecticut?

8. By the Delaware? Potomac? Savannah? Sabine?

9. By the Chattahoochee? Ohio? What States touch Lake Michigan?

10. What territories are bounded north by British America?

11. What border on the Pacific? What Territory is bounded by Mexico?

12. What territory South of Origan? West of Kansas? North of Texas?

· 13. Name the Gulfs and Bays in the State of Wis. Straits and Sounds.

14. Locate all the Capes and Islands. All the Mountains.

15. Name all the Lakes and principal rivers.

16. Which is the longest river in the U. S.? What is its length?

17. Which is the largest branch of the largest river? Next?

18. What falls in the U. S.?

19. Name the eastern branches of the Mississippi.

20. Name the western branches of the Mississippi.

NOTE.—See Map of U. S. for authority on all these questions.

21. Bound Maine. Describe the surface, soil and climate. Tell what it abounds in. Mention the chief pursuits of the people. Productions. The exports. What is said of Augusta? Eastport? Bangor? Bath? The Capital?

22. Bound and give a similar description of every state in the U. S.

23. Bound the United States as a whole.

24. Between what parallels of latitude are the U. S.?

25. What is the latitude of Albany? Madison, Wis.? New York? St. Louis? Boston? Washington? Chicago?

26. What state has the greatest amount of commerce?

27. Which is the greatest manufacturing state?

28. What is the largest city in the U. S.? In the Western States?

29. What are the principal Atlantic sea-ports?

30. Through what water would a vessel pass in going from St. Louis to New York? From Chicago to Boston?

31. What is the principal natural curiosity of the Middle States?

32. Bound the Eastern States as a whole and tell by whom and when they were settled, and describe in the same way.

33. Bound the Middle States as a whole.

34. Do the same with the Western States.

35. Bound the Southern States and tell when and by whom settled.

36. Give the general features of the Territories and tell how they are situated.

MEXICO AND THE WEST INDIES.

1. Bound Mexico. Give its latitude and longitude.

2. Give the principal features, climate and productions.

3. Give the latitude of Cuba. Hayti. P. 48.

4. Which is the largest of the West India Islands? P. 48.

5. What group north-east of it? P. 48.

6. What islands west of California? P. 48.

7. What mountain in Mexico? P. 48.

8. Give all the Bays and Gulfs in Mexico and West Indies. P. 48.

9. Name all the Islands and Capes.

10. Name all the Lakes and rivers.

11. What river between Mexico and W. States? P. 48.

12. Draw a map of Mexico. Also of the W. States.

MAP OF SOUTH AMERICA.

1. Bound South America.

2. How many square miles has it? Inhabitants?

3. What is its latitude? Longitude? In what zone does it lie?

4. How many states has South America? Name them. P. 85.

5. Bound each state, and give the capital of each. P. 85.

6. What states border on the Pacific? On the Atlantic? P. 85.

7. On the Caribbean Sea? What state has no sea coast? P. 85.

8. What state entirely west of the Andes. Which is the largest state? P. 85.

9. What ocean east of S. A.? West? What sea North?

10. Name all the Gulfs and Bays.

11. What Archipelagoes on the west coast of Patagonia? P. 85.

12. Locate and name all the Islands, Capes, Mountains, Lakes, and Rivers, and name the largest river and its branches. P. 85.

13. What is S. A.? How divided? For what distinguished? P. 87.

14. What is the climate? Soil? What are the products?

15. What plants are found in their native state? What can you say of the minerals? P. 87.

16. Wild Animals? Of the discovery and settlement of South America? P. 87.

17. What is the general form of S. A.? P. 95.

What range of mountains extend through the whole length? P. 95.

18. What countries of S. A. are crossed by the equator? P. 95.

19. Which division has the coldest climate?

20. Where is gold found? Silver? Copper? Mercury? P. 95.

21. Which is the largest city of S. A.?

22. How would you go from New York to Rio Janeiro? P. 95.

23. How do people travel in New Granada? P. 95.

24. What divisions have a temperate climate? P. 95.

25. Draw a map of South America.

EUROPE.

1. Bound Europe. How many square miles has it? P. 102.

2. How many inhabitants? What is its latitude? Longitude? P. 102.

3. In what zone is it? What are the principal divisions? P. 102.

4. Which is the most Northern? Eastern? Southern? Western?

5. Bound Norway and tell how many square miles it has. Inhabitants. Climate. Soil. Products. Government. Religion. Education. Principal towns. Mines. P. 104.

6. In a similar way describe and bound each of the divisions of Europe, and give the capitals. P. 102.

7. What ocean west of Europe? North? What sea North of Russia? P. 102.

8. What sea North of Prussia? East of Great Britain? West?

9. What sea separates Europe from Africa? P. 102.

10. Locate and name the balance of the seas. Bays. Gulfs.

11. Name the straits and channels, and tell what they connect and separate. P. 102.

12. Name all the islands and tell what government they belong to and what direction they are from the government to which they belong. P. 102.

13. Name all the Peninsulas and tell how situated. P. 102.

14. Give all the mountains, their situation and general direction. P. 102.

15. Give the name and locate all the Lakes. P. 102.

16. Give the names, sources and mouths of all the rivers and their general direction.

17. Name the largest river and its branches. P. 102.

18. What is the rank of Europe among the grand divisions? P. 104.

19. What is said of its coast and surface? Climate? Soil? Minerals? P. 104.

20. What is said of the animals? Agriculture? Population? Education? P. 104.

21. What is the condition of the people? What religion prevails? P. 104.

22. What divisions of Europe are entirely separate from the continent? P. 125.

23. What is the latitude of London? Paris? Rome? Lisbon? P. 125.

24. What countries of Europe are mountainous?

25. What is the climate of England? P. 125.

26. How does the climate of England compare with the countries of the same latitude on the Western Hemisphere? P. 125.

27. Which is the largest city? Describe it.

28. Which are the most commercial cities of Europe? P. 125.

29. Which are the greatest manufacturing cities? P. 125.

30. Draw a complete map of Europe.

31. How does the number of miles of coast line in Europe compare with that of the other grand divisions? P. 125.

32. What countries of Europe are Roman Catholic?

33. What countries are Protestant? P. 125.

34. What is the form of government of each country? P. 125.

ASIA.

1. Bound Asia. What is the number of its square miles? Number of inhabitants? P. 130.

2. What is its Longitude? Latitude? P. 130.

3. In what zones is Asia?

4. With what grand division is it connected? P. 130.

5. Name its political divisions. Their capitals. P. 130.

6. What division extends farthest North? P. 130.

7. Which farthest South? East? West?

8. What divisions have no sea coast? P. 130.

9. Name and locate the Oceans, Seas, Gulfs and Bays.

10. What is there remarkable in respect to the Caspian and Aral seas?

Ans.—They have inlets but no outlets.

11. Name the straits and channels and tell what they *connect* and what they *separate*.

12. Name the principal islands and tell where they are situated.

13. Give the peninsulas and principal capes of Asia. P. 130.

14. Describe the mountain scenery and deserts. P. 130.

15. Name and locate the Lakes, and give the sources, courses and mouths of the principal rivers in Asia. P. 130.

16. What is the size of Asia? Mention its natural characteristics. P. 132.

17. For what is it distinguished? What are cultivated? P. 132.

18. In what is it rich? What animals are found? P. 132.

19. Who inhabit it? What is their character? P. 132.

20. What did Asia formerly contain? P. 132.

21. What is said of agriculture and the arts? Religion? P. 132.

22. What has transpired here? P. 132.

23. What can you say of the great Chinese wall? P. 132.

24. Where and how high are the Himalaya mountains? P. 141.

25. Where is Mount Sinai? Ararat? P. 141.

26. What is the latitude of Calcutta? Pekin? Mecca?

27. How is Singapore situated? Describe it. P. 141.

28. What remarkable tree is found in Hindostan? P. 141.

29. In what have the Hindoos excelled? P. 141.

30. How would you sail from New York to Singapore? P. 141.

31. From Calcutta to Constantinople? P. 141.

32. What is the most noted product of China? P. 141.

33. Draw a *full Map* of Asia.

AFRICA.

1. Bound Africa. In what zones is it situated? P. 145.

2. Point out the principal divisions on the map. P. 145.

3. What states border on the Mediterranean Sea? Red?

4. On the Indian Ocean? Atlantic? P. 145.

5. What states have no sea coast? What are crossed by the equator?

6. What by the Tropic of Cancer? Tropic of Capricorn? P. 145.

7. What ocean east of Africa? West? What sea north? P. 145.

8. What ocean between Africa and Asia? P. 145.

9. Name and locate the principal Seas, Gulf and Bays? P. 145.

10. What strait at the entrance of the Mediterranean Sea? P. 145.

11. What channel between Mozambique and Madagascar? P. 145.

12. Name and tell how situated, the Islands, Isthmus and Capes. P. 145.

13. Describe the Mountains, Deserts and Oases.

14. Name the *lakes*, and give the source, course and mouths of the chief rivers of Africa. P. 145.

15. What is the position of Africa? What is said of its coast? P. 147.

16. What is the size of Africa? Climate? Minerals? P. 147.

17. What is said of its mountains and deserts? P. 147.

18. Of its soil and productions? Animals? Birds? P. 147.

19. By whom is Africa inhabited? What did Africa formerly contain? P. 147.

20. Name the Barbary States. How situated? P. 148.

21. What is said of the valley of the Nile? Climate?

22. What was Egypt anciently? Thebes? Cairo? P. 149.

23. Describe the pyramid near Cairo. P. 149.

24. Where are the Snow mountains? P. 155.

25. With what other grand division is Africa connected? P. 155.

26. What great desert in Africa? Describe it. P. 155.

27. What rivers in Africa? P. 155.

28. What is the government of Egypt? P. 155.

29. What states in the south temperate zone? P. 155.

30. Draw a complete map of Africa. P. 155.

31. For what is St. Helena noted? P. 155.

MAP OF OCEANICA.

1. In what two oceans are the Islands of Oceanica situated? P. 159.

2. Which are the three great divisions of Oceanica? P. 159.

3. What part of Oceanica constitute Malaysia?

4. What part Australasia? Polynesia? P. 159.

5. Name and locate the principal Islands. P. 159.

6. Give the Seas, Gulfs, Bays and Straits.

7. Mention the Capes, Mountains and Rivers. P. 159.

8. Which is the largest of the Sandwich Islands? P. 159.

9. Mention the principal towns.

10. Which is the largest town on the map of Oceanica? P. 159.

11. What division of Oceanica is nearest America? P. 162.

12. On what island is Mount Ophir? Where is Botany Bay? P. 163.

13. What islands are crossed by the equator? P. 163.

14. What is the latitude of the Sandwich Islands? P. 163.

15. What is the longitude of the Sandwich Islands? P. 163.

CHAPTER IV.

PHYSICAL GEOGRAPHY.

1. Of what does Physical Geography treat? P. 167. (Camp's High School Geography.)

2. What has Descriptive Geography taught of the surface of the earth? P. 167.

3. What is said of the Heat of the earth? (Geology) P. 167.

4. In what state is the interior of the earth supposed to be? P. 167.

5. What is said of the crust? How far has it been penetrated? P. 167.

6. How much of its composition is supposed to be known? P. 167.

7. How many simple elements are there? P. 167.

8. How many species of minerals are known? P. 167.

9. Which are the most common minerals? P. 167.

10. How are they combined? What is meant by the term rock? P. 167.

11. How are rocks classified? What are stratified rocks? P. 167.

12. Unstratified rocks? Igneous rocks? Plutonic rocks? P. 167.

13. Volcanic rocks? Aqueous rocks? Metamorphic rocks? P. 168.

14. Fossiliferous rocks? Describe the Granite rocks. P. 168.

15. What are the lowest stratified rocks? P. 168.

16. Mention the succeeding series in their order. P. 168.

17. What other unstratified rocks occur? P. 168.

18. What is said of the effect of certain rocks? P. 168.

19. Of changes now taking place? P. 168.

20. What are the Carboniferous rocks and what position found in? P. 168.

CONTINENTS AND ISLANDS.

1. What continent is the most simple in form? P. 169.

2. Is there any similarity in the two continents? What? P. 169.

3. Describe the characteristics of the Western Continent. Eastern. P. 169.

4. How are the islands divided?

5. What is said of the Continental islands? Volcanic? P. 169.

6. Of Coral islands? Coral Reefs? P. 170.

MOUNTAINS AND VOLCANOS.

1. How do Mountains occur? Volcanos? P. 170.

2. What is a Mountain system? To what do they correspond?

3. Describe the principal systems of the Western continent.

4. Of the Eastern continent. What is peculiar to each? P. 170.

5. What is meant by volcanic action?

6. How many volcanos are now active? P. 170.

7. Describe the chief volcanic regions. P. 170.

8. For what is Stromboli noted? Hecla? Cotopaxi? P. 171.

9. Jorullo? What are earthquakes?

10. Mention some of the most destructive. P. 171.

PLATEAUS AND PLAINS.

1. What are Plateaus? Plains? P. 171.

2. Which continent is distinguished for its plains?

3. Its plateaus? Describe the Plateau of N. America. P. 172.

4. Of S. America. Asia. Describe the Plain of N. America. P. 172.

5. Of S. America. Of Europe. Asia. Africa. P. 172.

WATER.

1. How are the waters of the land found? P. 172.

2. Describe Springs. Hot and Boiling Springs. P. 172.
3. What is meant by the basin of a river? P. 173.
4. Mention the principal basins on the land.
5. Upon what does the velocity of a river depend? P. 172.
6. In what direction do the largest rivers flow?
7. What are deltas? How formed? P. 173.

OCEAN.

1. What is meant by the basin of an ocean? P. 174.
2. Describe the Hydrographic system of the ocean. P. 174.
3. Describe the Waves, Tides, Currents. P. 174.
4. What causes the Antarctic Current? The Gulf Stream?
P. 174.
5. What causes the Equatorial Current? P. 175.
6. Give the cause of the South Connecting Current.
7. Point the course of the Japan Current. P. 175.
8. What is the cause of the Sargasso Sea? P. 175.

ATMOSPHERE AND MOISTURE.

1. What is the Atmosphere? What is Temperature? P. 175.
2. Upon what does the Temperature of a place depend? P. 175.
3. What causes wind? Hot winds? Hurricanes? P. 176.
4. What *causes* the *Trade Winds?* Return Trades? P. 176.
5. What is the cause of the Monsoons? Water Spouts? P.
176.
6. What is Dew? Frost? Mists and Fogs? Rain? Hail
and Snow? Give the cause of each. P. 177.
7. Where are rains periodical? Frequent? No rain? P. 177.
8. What is the annual fall of rain? What is meant by snow
line?
9. What are Glaciers? Climate? Isothermal Lines? P. 177.
10. Where is there the most rain, and how much?
Ans.—Cherrafongi, Southern India, $610\frac{3}{10}$ inches.
11. Where is the least rain, and how much is there?
Ans.—Fort Yuma, California, $1\frac{73}{100}$ inches.
12. By what is climate effected? P. 177.

PLANTS AND ANIMALS.

1. What is the estimated number of species of plants on the globe?　P. 178.

2. What is said of the native region of plants?　P. 178.

3. Of the distribution of plants?　P. 178.

4. What are most important influences effecting vegetation?

5. What is said of the vegetation of the Torrid Zone?　P. 178.

6. Of the Temperate Zone?　Frigid Zone?　P. 178.

7. What can you say of local and restricted botanical regions?　P. 178.

8. How are Animals adapted to different climates?　P. 179.

9. What is said of the clothing of animals of different zones?　P. 179.

10. What animals does the Torrid Zone contain?　P. 179.

11. What kind of birds?　Where are the coral tribes found?　P. 179.

12. What classes of animals belong to the different zones?　P. 179.

13. What is said of animals peculiar to particular regions?　P. 179.

14. How does Physical Geography differ from Political Geography?　P. 179.

15. Draw a new map of the U. S. with Isothermal Lines.

Note.—I could have multiplied these questions four fold: but perhaps the answers given would not have developed more principles than clear answers to the comprehensive questions already proposed.

CHAPTER V.
PENMANSHIP.

Remark.—The law requires that candidates for teaching shall be examined in Penmanship. Yet with few exceptions little or no attention is given to the subject; as a result, poor penmanship is generally found in all our schools. Good penmanship is an exception, unless a teacher has been employed especially for that branch. This is wholly unnecessary.

The following references to questions on penmanship are to the "MANUAL OF PENMANSHIP," by Payson, Dunton, Scribner & Hays. Authors of the COMBINED SYSTEM OF PENMANSHIP.

P. stands for the page of the Manual.

1. What is Penmanship?

Ans.—*The art of writing.*

2. What may reasonably be expected from the school course of writing? P. 19.

3. Which should be taught first, knowledge of forms, or command of the pen? P. 21.

4. Can a bad writer teach penmanship successfully? P. 23.

5. Mention all the requisites for an exercise in writing. P. 25.

6. How many are there in all?

7. What is said about "*Script and Print?*" P. 37.

8. Why should one differ in form from the other?

9. What can you say about *Lines* and *Angles?* P. 42.

10. What degrees should the Angles be? P. 45.

11. What can you say of the *Position? Rests* and *Movements?* P. 46.

12. What can you say of the *Elements* and *Principles?* P. 57.

13. What is the first *Element?* P. 60.

14. What is the second *Element?* P. 60.

15. Describe the third *Element.* Fourth. Fifth. P. 60.

16. How many principles from the five Elements? P. 60.

17. The main lines have a slope of how many degrees? P. 60.

18. The connecting lines how many degrees? P. 60.

19. How many Elements are there in the First Principle? P. 60.

20. How many Elements form the Second Principle? P. 61.

21. How many form the Third Principle? P. 61.

22. How many form the Fourth Principle? P. 61.

23. How many form the Fifth Principle? P. 62.

24. Explain the Sixth Principle.

25. What Elements are there in the letter *O?* (See elements Plate.)

26. What is said about the *Scale of Length?* P. 63.

27. Give the rule for *Scale of Length.* P. 63.

28. Give the rules for small letters. P. 65, 66.

29. Give the caution mentioned on page 72.

30. Give the analysis of the figures. P. 74.

31. Describe the principles found in the Capitals. P. 76.

32. Give the general rule for *Capital Letters.* P. 78.

33. Give the caution mentioned on pages 79 and 80.

34. What is said about the classes of small letters? P. 87.

35. What letters belong to the *First Class?* P. 87.

36. To the Second Class, Third Class and Fourth Class? P. 87.

37. What are the characteristics of the letters? P. 88.

38. What is said about the order in which the small letters are introduced? P. 88.

39. Describe the combination of letters, and give the rules. P. 94.

40. Give and explain the *Schedule of Topics* and *Course of instruction* as found on pages 104 and 105. Explain topics on page 146. Mention anything else important in teaching penmanship.

CHAPTER VI.

BOOK-KEEPING.

Although the *Law* does not demand that a candidate shall be examined in Book-Keeping, yet that is no reason why the teacher should not be *qualified* to teach that which all the *youth* of the *land* are *going* forth to practice. A few questions are therefore given on this important branch.

The following references in the questions on *Book-Keeping* are to "BRYANT & STRATTON'S NATIONAL BOOK-KEEPING," "High School Edition." The best work published on the subject.

P. stands for page.

1. What is Book-Keeping? In what does business consist? P. 11.

2. How many methods of Book-Keeping are there, and how distinguished? P. 11.

3. Which is the better method, and for what reason? P. 11.

4. What is the characteristic feature of Double Entry? P. 11.

5. Why must each Transaction be entered twice on the Ledger? P. 11.

6. What are the three main books in Double Entry? P. 11.

7. Which two are sometimes combined in one? P. 11.

8. Describe the Day Book. What should be the character of its expression? P. 11.

9. Why is the Day Book alone produced in Courts? P. 11.

10. What other importance has it? Describe the Ledger. P. 12.

11. What is the character of the Ledger? P. 12.
What is an account? P. 12.

12. Which is the Debit and which the Credit side of an account? P. 12.

13. What is a Resource? Liability? Cash Term? P. 14.

14. For what is cash Account Debited? Credited? P. 14.

15. What is shown by the difference between the sides? P. 14.

16. Which side of a Cash account must be the greatest, and why? P. 14.

17. What is meant by Bills Receivable? Payable? P. 14.

18. For what is Bills Receivable Account Debited? Credited? P. 14.

19. For what is Bills Payable Account Debited? Credited? P. 14.

20. What is shown by the difference? P. 14.

21. What is a Merchandise Account? With what Debited? Credited? P. 14.

22. What is Real Estate? How is the Account kept? P. 15.

23. Wherein are Accounts with Bank Stock, Railroad Stock, &c., similar to Merchandise? P. 15.

24. What are Personal accounts? With what debited? Credited? P. 15.

25. What is meant by Shipment or Adventure? How do they differ from Merchandise? What is meant by the term Stock? P. 15.

26. What do Stock Accounts show? Explain the manner of keeping Stock Accounts. P. 16, and FORMULA (P. 17.)
State the six general principles in keeping accounts. P. 16.

27. What should the pupil keep in view when Journalizing? P. 20.

28. What is Posting? Why is it necessary to observe care in Posting? P. 20.

29. When should the Check Mark (v) be made in the Day Book, and where? P. 20.

30. Why is a Trial Balance so called? P. 27.

31. Can a correct Trial Balance be had which will contain simply the *Balance* of the Ledger Account? P. 28.

32. What will a Ledger properly kept show at any time? P. 28.

33. What kind of Resources can not be shown from the Ledger? P. 28.

34. How many classes of Accounts are there? P. 29.

35. What Accounts are called Real? What Representative? P. 29.

36. For what purpose is Stock Opened? P. 32.

37. What is the chief difficulty with pupils in closing the Ledger? P. 33.

38. Describe the manner of closing the Ledger in regular steps. P. 34.

39. Define *Cash Book*. *Bill Book*. Commission Ledger Book. Invoice Book. Describe the form of Notes, Drafts, &c.

40. State any other principle in Book-Keeping not implied in the foregoing questions.

CHAPTER VII.
GRAMMAR.

REMARK.—Many candidates can *recite* the Grammar from beginning to end, yet in the application of the *principles* in analyzing language, they fail sadly. Again those who are able to tell the part of speech readily in the sentences taken from the Grammar, still make many bad mistakes in general analysis. It is of the highest importance, therefore, that every candidate should so thoroughly qualify himself in the principles of the science, that he may be able to judge of the part of speech from the OFFICE the word performs in the sentence. Every good student of Grammar knows that the SAME WORD varies its *part of speech* as it varies its *office* in different sentences.

NOTE.—The following references in the questions on Grammar are to Clark's English Grammar, Revised Edition. *Def.* stands for *definition*. *Obs.* for *observation*. *Rem.* for *remark*. *Pr.* for *principle*. *R.* for *rule*. *Ex.* for *example*. *P.* for *page*.

LANGUAGE.

1. What is Language? Def. 1.

2. How are *thoughts* and *feelings* indicated? Def. 1. Obs. 1.

3. What can you say of Natural Language? Def. 1. Obs. 2.

4. Describe Artificial Language. Def. 1. Obs. 3.

5. Of what does Spoken Language consist? Def.

6. Of what does Written Language consist? Def.

GRAMMAR.

1. What is Grammar? Define General Grammar. Def. 4. Obs. 1.

2. What is Particular Grammar? Def. 4. Obs. 2.

3. What should every Particular Grammar include? Def. 4. Rem.

4. What is English Grammar? Define a letter. Def. 5, 6.

5. Define a *Word*. What is a Phrase? Def. 7, 8. Rem.

6. Point out the Phrases in the following:

> "At midnight, in his guarded tent,
> The Turk was dreaming of the hour
> When Greece, her knee in suppliance bent,
> Should tremble at his power."

7. What is a Sentence? Def. 9.

8. Illustrate your definition by examples. Def. 9, Ex.

WORDS.

1. Words are distinguished as how many Parts of Speech? Prin.

2. Give the Parts of Speech. Prin.

3. Describe a *Noun*. A *Pronoun*. An Adjective. Def. 10–12.

4. Illustrate each of the above by an *example*.

5. Describe a Verb, and tell what belongs to it. Def. 13.

6. Define an Adverb. A Preposition. Def. 14, 15.

7. What is a Conjunction?

Ans.—A word used to connect words, phrases, and sentences.

8. What is an Exclamation? Def. 17.

9. What is a Word of Euphony? Def. 18.

10. Give examples of Words of Euphony.

"There are no idlers here." "Now, then, we are prepared to defend our position."

PHRASES.

1. For what are Phrases used? Rem. P. 19.
2. How are Phrases distinguished? Prin. P. 19.
3. What is a Substantive Phrase? Def. 19.
4. Point out and tell the offices the Phrases perform in the following sentences:

"To be, contents his natural desire." "His being a minister prevented his rising to civil power." "I doubted his having been a soldier."

5. What offices do Substantive Phrases perform? Def. 19, Obs.

6. What is the office of an Adjective Phrase? Def. 20.

7. Designate the Adjective Phrases in the following:

"The time of my departure is at hand." "Forgetting the things that are behind I press forward."

8. What is an Adverbial Phrase? Def. 21.

9. Designate the Adverbial Phrases in the following:

"God moves in a mysterious way." "Truth crushed to earth will rise again."

10. Define an Independent Phrase. Def. 22.

11. What office does an Independent Phrase perform in a sentence? Def. 22, Obs.

12. Designate the Independent Phrases in the following:

"The hour having arrived we commenced the exercises."
"The sun having risen, we set sail."
"The bugle having sounded, the charge was made."

13. If you approve of distinguishing Phrases according to their forms, instead of the offices which they perform, tell how many classes there are, and name them. Def. 22, Prin.

14. Describe a Prepositional Phrase. Give an Ex. Def. 23.

15. Describe an Infinitive Phrase. Give an Ex. Def. 24.

16. Describe a Participal Phrase. Give an Ex. Def. 25.

17. Give a sentence illustrating an Independent Phrase. Def. 26.

18. Of what does a Phrase consist? Def. 26, Pr.

19. What are the principal elements of a Phrase? Def. 27.

20. Define the Adjuncts of a Phrase. Def. 28.

21. Designate the Principal Elements and Adjuncts in the fol·lowing:

"Birds sang amid the whispering shade."
"Rays of limpid light gleamed round their path."

22. What is the leader or connective of a Phrase? Def. 29, Obs.

23. Point out the leaders or connectives in the following:

"Like a spirit it came, in the van of a storm."
"Enough remains of glimmering light
To guide the wanderer's steps aright."
"I am monarch of all I survey;
My right there is none to dispute."

24. Define a Participle. Describe a *Subsequent* or *essential element* of a Phrase.

25. When any element of a Phrase is suppressed, how do you treat that part of the Phrase which is expressed? P. 23. Obs. 2.

26. Designate the Subsequent of a Phrase, and illustrate the principle in question 25th from the following:

"At parting, too, there was a long ceremony in the Hall; buttoning up great coats, tying on woolen comforters, pinning silk handkerchiefs over the mouth and up to the cars, and grasping sturdy walking canes to support unsteady feet." "These crowd around to ask him of his health." "William came home." "Mary has come to school early." "I love to see the sun rise."

SENTENCES.

1. What is a Sentence? P. 23. Rem.

2. What are the Elements of a Sentence? Def. 32.

3. What is essential to the structure of a Sentence? Def. 32.

4. What is the Subject of a Sentence? Def. 32. Rem. 1.

5. What are the parts of a Sentence? P. 24. Prin.

6. What is the Predicate of a Sentence? Def. 32. Rem.

7. What are Principal Elements? Adjunct Elements? Def. 33, 34.

8. Point out the Principal and Adjunct Elements in the following sentences: ·

"The night passed away in song." "The King of Shadows loves a shining mark."

3

> "There in his noisy mansion, skilled to rule,
>   The village master taught his little school."

9. Define Subordinate Elements, and designate them in the following : *

"Lend me your songs, ye Nightingales." "Oh Liberty! I wait for thee." "There are no idlers here." "I sit me down a pensive hour to spend."

10. What *must* every Sentence have? Def. 35.

11. Is there any distinction between a Logical and Grammatical Subject? If so give it.

12. What is the difference between a Grammatical and Logical Predicate? Give the distinction between a Grammatical and Logical Object. Def. 37, 38.

13. In the following Sentences tell whether the Subject, Predicate and Object are Grammatical or Logical:

"Birds fly." "Knowledge is power."
"They that seek me early, shall find me."
"To do good, is the duty of all men."
"At what time he took orders, doth not appear."
"That all men are created equal is a self-evident truth."
"Thou art perched aloft on the beetling crag."
"I regret his being absent."
"The fool hath said in his heart, there is no God."
"God said, let there be light."
"God never meant that man should scale the heavens,
  By strides of human Wisdom."

DIAGRAMS.

1. What is a diagram, and what is its object in Grammar? P. 36.

2. What determines the position of an element in the Diagram? Rem.

3. Illustrate by an example. P. 36.

GENERAL RULES.

1. What position do the principal elements occupy in the Diagram? R. P. 36.

2. What position do the subject, predicate, and object, occupy in a Diagram? R. 2, 3, 4. P. 36, 37.

3. What position do the Adjunct, Conjunction and Relative Pronoun occupy in a diagram? R. 5–10, 12. P. 37, 38.

CLASSIFICATION OF SENTENCES.

1. How are *Sentences* distinguished? Pr. P. 38.

2. Describe an Intransitive Sentence, and give an example. Def. 43.

3. Define a Transitive Sentence, and give an example. Def. 44.

4. In the following Sentences, designate which are *Transitive*, and which Intransitive:

"God is love." "On some fond breast the parting soul relies." "Virtue secures happiness." "Fishes swim." "Industry promotes health and wealth." "John walks."

5. Define a *Simple Sentence*. A *Compound Sentence*. Def. 45, 46.

6. In the following, tell which are Compound, and which Simple:

Frank is diligent. James is quiet. The boys run. Maggie and Flora study Latin. "Temperance elevates and ennobles man."

7. What are the clauses of a Compound Sentence? Def. 46, (*b.*)

8. Designate in the following, what clauses are Compounded:

John and James study Philosophy. Anna reads Latin and French. Phebe studies and recites Algebra and Geometry. Slowly and sadly Minnie and Maggie ride and walk up yon high and distant mountain and woodland.

9. What is a Mixed Sentence? Def. 46, (*e.*)

10. Designate which of the following Sentences is Mixed:

"Time slept on flowers and lent his glass to hope."
"Who can observe the careful ant, and not provide for future want."

11. What is a Principal Sentence? Def. 47.

12. Describe an Auxiliary Sentence. Def. 48.

13. Define a Complex Sentence. Def. 48. Obs. P. 42.

14. Designate which of the following Sentences is Principal. Which Auxiliary. Which Complex:

a. Man is mortal.

b. He hath brought many prisoners from Vicksburg.

c. "The fur that warms a monarch warmed a bear."

d.   "Sweet was the sound, when oft, at evening's close,

e.   Up yonder hill the village murmur rose."

f. "I have a temple in every heart, that owns my influence."

15. How are Auxiliary Sentences distinguished? P. 42, Prin.

16. Define a Substantive Sentence. Def. 49.

17. What is an Adjective Sentence? Def. 50.

18. Describe an Adverbial Sentence. Def. 51.

19. In the following, designate the Substantive Sentences, the Adjective Sentences, and Adverbial Sentences:

a. "That man is liable to err, is evident to all."

b. "James refused to tell what caused him to commit the deed."

c. "He that getteth wisdom loveth his own soul."

d. "That life is long, which answers life's great end."

e. "Teachers are anxious that their pupils should improve."

f.   "How dear to my heart are the scenes of my childhood,

g.   When fond recollection presents them to view."

ETYMOLOGY.

1. Define Proximate Analysis. Ultimate Analysis. Rem. 2, 3, P. 69.

2. What is embraced in the Science of Language? Pr. P. 69.

3. Define Orthography. (See first Chap. of the Examiner.)

4. What does Etymology treat of? Syntax? P. 69.

5. What is Prosody, and of what does it treat?

6. What does a true system of Analysis require? Rem. P 69.

7. How are words distinguished? Pr. 70.   .

8. Define a Radical word. Derivative word. Def. 52, 53.

9. Must a word that is Radical in English, be Radical in the language from which it is derived? Obs. P. 70.

10. How are the parts of a compound word usually written? Obs. P. 70.

11. What is the Basis of a Compound word? Adjunct? Def. 56, 57.

12. What is the distinction between a Derivative and Compound word? R. P. 71.

13. Define a Prefix and its office.   Def. 58.
14. Describe a Suffix and its office.   Def. 59.
15. How are Prefixes and Suffixes distinguished?   Pr. P. 71.
16. What is a Separable Radical?   Def. 60.
17. Describe an Inseparable Radical.   Def. 61.
18. Designate the Radical and Derivative words in the following:

a. "The profoundest depths of man's intellect can be fathomed."
b. "In the loftiest flights of his imagination he can be followed."
c. "Dryden often surpasses expectation."
d. "Pope never falls below it."
e. "Behold the majestic monarch of the clouds."

19. Designate the Simple and Compound words in the following:

a. "Flag of the free heart's only home,
b. By angel hands to valor given,
c. Thy stars have lit the welkin dome,
d. And all thy hues were born in heaven.
e. Forever float that standard sheet;
f. Where breathes the foe but falls before us;
g. With Freedom's soil beneath our feet,
h. And Freedom's banner streaming o'er us."

20. Designate the Prefixes and Suffixes, the Separable and Inseparable Radicals in the following:

a. Goodness sake absolve dangerous formations.
b. Undertake to overtake consular uniforms.
c. Reform undelectable and convertable elections.
d. Reconstruct the comparable and indissoluble Union rightfully.

OF THE NOUN.

1. What is a Noun?   Describe its office.   Def. 62.
2. Give the order of parsing a Noun.
3. What is the difference between a Proper and a Common Noun?   Def. 63.
4. Illustrate the above by examples.
5. When a Noun denotes the quality of a thing, what kind of a Noun would you call it?   Def. 65.
6. Describe a Collective Noun.   Def. 66.

7. Define a Verbal Noun.   Def. 67.

8. Is it essential to the Science of Grammar that Nouns should be classified as Common and Proper?

Ans.—It is not.

9. May the office of a Substantive be performed by Words, Phrases, and Sentences?   Obs. 4.   P. 74.

10. May a word that is usually a noun perform the office of an Adjective, Adverb, or Verb?   Obs. 5.   P. 75.

11. May the same word perform the office of any part of speech?

12. Designate the Nouns in the following:

a. An Iron fence.
b. Go home and come back.
c. But if you mouth it.
d. The good alone are great.
e. I grudge thee not the much or the little thou mayest receive.
f. "'Tis Heaven itself that points out an hereafter."
g. "Your if is the only peace-maker; much virtue is in if."
h. "They came down with hark, and whoop, and wild balloo."

MODIFICATION OF NOUNS.

1. How many Genders have Nouns?   Name them.   Rem. P. 75.

2. What is Person as belonging to Nouns?   How many Persons are there?

3. What belong to Nouns?   Pr. P. 76.

4. What does the Masculine Gender denote?   Def. 69.

5. What Nouns are of the Feminine Gender?   Def. 70.

6. What Nouns are of the Neuter Gender?   Def. 71.

7. What Gender do you apply to the names of infants and young animals?   Obs. 2.   P. 76.

8. What Gender do you apply to the names of objects personified?   Obs. 3.   P. 76.

9. What Gender do you apply to the following words: Parent, Cousin, Friend, Bird, Sheep, Deer?   Obs. 4.   P. 76.

10. Give the feminine of *actor, author, host, hero, jew, lion, bear, tiger, man.*

11. Give the masculine of *governess, princess, lady.*

PERSON.

1. Define Person as belonging to Nouns.
2. Give an example of a Noun of the First Person.
3. What Nouns are of the Second Person? Def. 73.
4. What Nouns are of the Third Person? Def. 74.

NUMBER.

1. Define Number as belonging to Nouns.
2. What Nouns are of the Singular Number? Def. 75.
3. What Nouns are of the Plural Number? Def. 76.
4. How do you determine the Number of Nouns? Obs. P. 78.
5. How do you form the Plural of Nouns whose Singular ends in s, ss, sh, x, ch, &c. Obs. 2. P. 79.
6. How do Nouns ending in y, form their Plural? Obs. P. 79.
7. How do you form the Plural of Nouns ending in f final? Obs. 4. P. 79.
8. Nouns ending in fe form their Plural in what manner? Obs. 5. P. 79.
9. Give the plural of Book. Pen. Gas. Lynx. Hero. Lady. Folly. City. Beef. Loaf. Wolf. Wife. Child. Man. Ox. Foot. Mouse.
10. How do Compound Words form their Plural? Obs. 7. P. 80.
11. How do you form the Plural of Nouns, having titles prefixed or annexed? Obs. 9. P. 80.
12. Give the Plural of the following: Inkstand. Race-horse. Father-in-law. Arm-full. Ignis-fatuus. Miss Brown. Doctor Smith. Gold.
13. Give the Singular of Tongs. Vespers. Literati. Scissors.
14. Give the Number of the following: News. Wages. Sheep. Horse. Foot. Cattle.
15. Give the Plural of Genus. Index. Axis.

CASE.

1. What does Case in Grammar mean? Rem. 2. P. 82.

2. How many Cases do Nouns have? Name them. Pr. P. 82.

3. Define the Nominative and its office. Def. 77.

4. In what Case is the Subject of a Sentence? Obs. P. 83.

5. Define the Possessive Case and its office. Def. 78.

6. Does the Possessive Case always denote ownership or possession? Ex.—Childrens' Shoes are high. Obs. 5. P. 84.

7. Do Nouns in the Possessive Case more frequently perform the office of Nouns than of Adjectives? Obs. 7. P. 84.

8. What can you say of the Objective Case? Def. 79.

9. When is a Noun or Pronoun in the Independent Case? Def. 80.

10. Do Nouns change their forms to represent their Cases? Obs. 7. P. 89.

11. Analyze and parse the words in *Italics:* My *Book is* new; *John's is* old. *Mine is* little used; *Yours is* soiled. John is a friend of mine. Does the word *mine* mean *my friend?* Test it by this example: John is my enemy; but he is a friend of "my friend." Ex. P. 86.

PRONOUNS.

1. Define a Pronoun and its office. Def. 81.

2. What do you mean by Antecedent? Obs. 1. P. 88.

3. Give the Antecedents in the following:

John is sick; I saw him yesterday. I am glad that Charles has received a good education, it is what few poor boys have the perseverence to accomplish.

4. What does *it* personate in the following sentences?

It is I. It is he. It is she. Who is it? It is they. It is the men. It is the children. It is the women.

5. How many kinds of Pronouns are there? Pr. P. 88.

6. Define a Personal Pronoun. Def. 82.

7. How many Personal Pronouns are there?

8. Decline the Personal Pronouns. P. 89.

9. What belong to Pronouns? Pr. P. 89.

10. How many Pronouns have a special form to denote sex?

11. Define the Relative Pronoun and its office. Def. 83.

12. Which of the Relative Pronouns are declined to indicate the Cases? Def. 83. Obs. 3.

13. When should we use who? When which? When that? P. 92.

14. What is there peculiar about the word *what?* Obs. 7. P. 92.

15. What can you say of the words whoever, whichever, &c.? Obs. 8. P. 92.

16. Parse *as* and *than* in the following:

"Such as I have, give I unto thee." "We have more than heart could wish."

17. What is an Interrogative Pronoun? Def. 84.

18. Describe an Adjective Pronoun. Def. 85.

19. How many distinct offices does every Adjective Pronoun perform? Obs. 2. P. 94.

20. Analyze and parse the following:

That that, that that little fellow mentioned, is such a troublesome *that,* that it might be considered a very mischievous little *that.*

ADJECTIVES.

1. Define an Adjective, and describe its office. Def. 86.

2. How are Adjectives distinguished? Pr. P. 97.

3. What is a Qualifying Adjective? Def. 87.

4. Define a Specifying Adjective. Def. 88.

5. What are Proper Adjectives? Obs. 1. P. 98.

6. What are Interrogative Adjectives? Obs. 2. P. 98.

7. How many classes of Specifying Adjectives are there? Pr. P. 98.

8. Define a Pure Adjective. Def. 89.

9. Describe a Possessive Adjective. Def. 90.

10. What are Possessive Adjectives derived from. Note, P. 99.

11. What are Numeral Adjectives? Def. 91.

12. Define a Verbal Adjective. Def. 92.

13. How are Verbal Adjectives distinguished? Pr. P. 100.

14. How many Degrees of Comparison have Adjectives? Pr. P. 101.

15. What does the Diminutive denote? Def. 93.

16. Define the Positive Degree. The Comparative Degree. Def. 94, 95.

17. What does the Superlative Degree express? Def. 96.

18. How are Adjectives compared? Obs. P. 102.

19. Compare three Adjectives by using *er* and *est.*

20. Compare three Adjectives by using *more* and *most.*

21. Compare three by using *less* and *least.*

22. Compare three that are Irregular.

23 Compare *after, top, round, square, triangular, infinite.*

24. Designate which of the following Adjectives are Qualifying, Specifying, Verbal. Which can be Compared. Which are Pure, Numeral, Possessive, Transitive, Intransitive: *Able, bold, eager, good, honest, that, three, infinite. several, standing, loving, unknown. Give reasons* for all your statements. Exercises, P. 103.

25. What are Adjective Phrases and Sentences? Rem. P. 106.

26. Give examples, with reasons. P. 106.

VERBS.

1. Define a Verb. How many kinds of Verbs are there? Def. 97.

2. What is a Transitive Verb? Def. 98.

3. What is an Intransitive Verb? Def. 99.

4. What is a Neuter Verb? Def. 100.

5. Form Sentences containing each kind of Verb.

6. How many Voices have Verbs? Pr. P. 108.

7. Define an Active Verb. Def. 101.

8. Describe a Passive Verb. Def. 102.

9. Give examples of each Voice.

10. May Intransitive Verbs take the Passive form? Obs. 5. P. 109.

11. Give an example of an Intransitive Verb in the Passive Voice. Obs. 5. P. 109.

MODE.

1. Define Mode, and tell how many Modes Verbs have. Pr. P. 109.

2. What does the Indicative Mode indicate? Def. 103.

3. What does the Potential Mode indicate? Def. 104.

4. What are the signs of the Potential Mode? Def. 104.

5. What does the Subjunctive Mode indicate? Def. 105.

6. What Conjunctions are commonly used with the Subjunctive Mode? Def. 105.

7. For what is the Imperative Mode used? Def. 106.

8. What Person and Number is the Subject of the Verb in the Imperative, and what Tense is the Verb? Def. 106. Obs

9. When is a Verb in the Infinitive Mode? Def. 107.

10. What is the sign of the Infinitive Mode? Def. 107. Obs.

11. After what Verb is the sign *to* omitted? Def. 107. Obs. 2.

12. Give Sentences with Verbs in each of the Modes.

PARTICIPLES.

1. What is a Participle? From what is the word derived? Def. 108.

2. Describe a Simple Participle. Def. 109.

3. What is a Compound Participle? Def. 110.

4. Define a Present Participle. Def. 111.

5. Describe a Past Participle. Def. 112.

6. Which Participle is Active? Def. 112. Obs. 3.

7. Which Participle is Passive? Def. 112. Obs. 4.

8. Are Participles ever used as any other parts of speech? Obs. P. 115.

9. Write Sentences containing each of the Participles.

10. What can you say of Participles being Transitive or Intransitive? Obs. 9. P. 114.

TENSE.

1. What does Tense mean? Def. 113.

2. How many kinds of Tense have we? Def. 113. Rem.

3. How many Tenses have Verbs? Name them. Pr. P. 115.

4. Define the Prior Past Tense or Pluperfect. Def. 114.

5. What is the sign of the Prior Past Tense? Def. 114. Obs.

6. Define the Past Tense. Give an example. Def. 115.

7. What is the sign of the Past Tense? Def. 115. Obs.

8. Define the Prior Present, or Perfect Tense. Def. 116.

9. What is the sign of this Tense? Def. 116. Obs. 3.

10. What does the Present Tense denote? Def. 117.

11. Has this Tense a sign? What is it? Def. 117. Obs. 1.

12. What does the Prior Future Tense denote? Def. 118.

13. Give the sign of this Tense.

Ans.—Shall have, or will have.

14. What does the Future Tense denote? Def. 119.

15. What is the sign of this Tense? Def. 119. Obs.

16. What is said about *shall* and *will?* Def. 119. Obs.

17. Correct the following where it is necessary:

I will be drowned. No body shall help me.

18. Write sentences in each of the Modes and Tenses.

19. Analyze and Parse the following, giving reasons:

"Could I forget what I have been, I might the better bear what I am destined to."

CONJUGATION OF VERBS.

1. What do you understand by the Conjugation of a Verb? Rem. P. 120.

2. What is a Regular Verb? An Irregular Verb? Def. 120, 121.

3. How does the Regular Verb form its Past Tense, and Past Participle? Def. 120.

4. What is a defective Verb? Def. 122.

5. Define an Auxiliary Verb. Give an example. Def. 123.

6. What Verbs are sometimes Auxiliary and sometimes Principal? P. 121.

7. Give examples of each. Def. 123. Obs.

8. Conjugate the Irregular Verb *Be.* P. 124.

9. Give a Synopsis of the Verb Study, by writing the First Person Singular in each of the Modes and Tenses. P. 126.

10. Write out the Paradigm of the Verb *"Press"* in all its Modes, Tenses, Persons, and Numbers. P. 130.

11. Do the same with the verbs *see, lie, say.* P. 136.

12. Write out the Present and Past Tenses, and Present and Past Participles of the following Verbs:

a. Arm, arise, bear, begin, beset, bet, bid, bite, build, chide, come,

b. cast, crow, cost, do, dare, dig, draw, dress, drive, eat, fight, flee,

c. fly, forget, go, hang, hear, hide, hit, hold, hurt, lade, leak, let,

d. mode, mean, outdo, pen, put, read, rid, rise, run, say, see, sell,

e. set, sit, short, shine, smite, spill, spit, stove, stay, stick, swear,

f. swim, teach, tell, tread, wax, weep, wet, write.

13. Define a Unipersonal Verb. Def. 124.

14. Give examples to illustrate Unipersonal Verbs.

15. Analyze and parse the following:

"*Methinks it was Clara.*"

ADVERBS.

1. Define an Adverb and its office. Def. 125.

2. What may Adverbs consist of? Def. 125. Obs. 4.

3. Write Sentences in which a *word* is an Adverb. In which a Phrase is an Adverb. In which a Sentence is an Adverb.

4. Into how many classes are Adverbs divided? Pr. P. 151.

5. Define an Adverb of Manner. Def. 126.

6. What is an Adverb of Circumstance? Def. 127.

7. Define Adverbs of Time and Place. Def. 127. Rem. and Obs. 2.

8. How are Adverbs Modified? Prin. P. 158.

9. Compare *Soon. Wisely. Quickly.*

10. Parse the following:

a. "*From crag to crag they passed.*"

b. In a moment he flew quickly past.

c. How is it possible not to feel a profound sense of the responsibleness of this Republic to all future ages.

PREPOSITIONS.

1. Define a Preposition and its office. Def. 128.

2. By what is a Preposition always followed?

ANS.—By a Noun or Pronoun, Phrase or Sentence.

3. What is the Consequent term of relation in the following:

a. The man of God was there. He stood before the people.

b. Time spent in receiving impertinent visits is wasted.

c. And cries of "Live forever," struck the skies.

d.   It is not fit for such as us
           To sit with rulers of the land.

4. Analyze and parse the following:

To him, who in the love of Nature, holds communion with her visible forms, she speaks a varied language.

CONJUNCTIONS.

1. Define a *Conjunction* and give its office.
2. How many kinds of Conjunctions are there?
3. Do any other parts of speech than Conjunctions perform the office of a *Conjunction?* Obs. 6.   P. 164.
4. What can you say of Relative Pronouns, as connectives?
5. Example:

> "The grave, that never spake before,
>   Hath found at length, a tongue to chide."

EXCLAMATIONS.

1. Define an Exclamation.   Def. 130.
2. Of what may Exclamations consist?   Obs. 1.   P. 165.
3. By what are Exclamations or Interjections followed?   Obs. P. 165.
4. Give an example to illustrate Exclamations.
5. Example:

O, for a lodge in some vast wilderness!

WORDS OF EUPHONY.

1. Define a Word of Euphony.   Def. 131.
2. What is the office of a Word of Euphony?   Obs. 1.   P. 166.
3. Give an example to illustrate the office of such words.
4. Analyze and parse the following:

"There are no idlers here."

5. Also the following:

> "I sit me down a pensive hour to spend."
> "His teeth, they chatter, chatter still."

SYNTAX.

1. Define Syntax.   Def. 132.
2. Define a Sentence, and tell of what it consists.   Def. 132.
3. What are the Principal Elements of a Sentence?   Def. 132.
4. What are the Adjuncts of a Sentence?   Def. 132.
5. What is the Rule for the Subject of a Sentence?   R. 1.
6. State the Rule that requires a Verb to agree with its subject.   R. 2.
7. Is it true that Verbs agree with their subjects in Number and Person?   R. 2.   Rem.
8. What Number must the verb be in when two or more singular Subjects are taken together?   Note 3.   P. 196.
9. A Collective Noun indicating unity, requires the verb to be in what Number?   Note 4.
10. A collective Noun indicating Plurality, requires the verb to be in what Number?   Note 7.
11. When there are two or more Subjects taken separately and differing in Person, how many Verbs should there be, and how varied?   Note 8.
12. What Mode and Tense of a Verb should always be used?   Note 9.
13. Correct the following, if necessary:

a. "The rapidity of his movements were beyond example."
b. "To study Mathematics require maturity of mind."
c. "Wisdom and folly govern us."
d. "An effort is making to abolish the war."
e. "They don't ought to do it."
f. Had I known the character of the lecture, I would not have went.

14. What is the Rule for the object of an action or relation?   R. 3.
15. What may the Object of a Verb be?
Ans.—A Word, Phrase, or Sentence.
16. How many Rules are required to parse Pronouns?
17. Give them, and give examples.
18. Give the Rule for the Independent Case.   R. 6.
19. State the Rule for Adjectives.   R. 7.
20. Give the Rule for Specifying Adjectives.   R. 8.

21. Correct the following, and give the reasons:

a. William feels badly to night, and I feel sadly.
b. How beautifully it looks. Note 4. P. 250.

22. Where should an Adjective in Predicate be placed? Note 5. P. 251.

23. Give the Rule for Adverbs. R. 9.

24. What should be the position of Adverbs? Note 2. P. 258.

25. State the *Rules, Notes* and *Obs.* that apply to Participles. R. 10.

26. Give the *Rule, Obs.* and *Notes* pertaining to the verb in the Infinitive.

27. What is the **Rule** for Prepositions? **R. 12.**

28. Give the important Obs. and Notes under the Rule.

29. State the Rule and Obs. for Conjunctions. R. 13.

30. What is the Rule for Exclamations? R. 14.

31. State the Principal in regard to Words of Euphony. Pr. P. 279.

32. What is the position of Words of Euphony? Note, P. 279.

33. Give the five General Rules for constructing Sentences. P. 280.

34. Correct the following where it is necessary:

a. "The bill passed the Lords' house, but failed in the Commons."
b. "It is curious enough that this Sentence of the bishop's, is itself un-grammatical."
c. "We should presently be sensible of the melody suffering."
d. "Heaven opened widely her everlasting gates."
e. · James feels very badly about it.
f. The sight appeared terribly to me.
g. He mentions Newtons writing of a Commentary.
h. In the choice they had made of him for restoring of order.
i. To prevent it bursting out with open violence.
j. They refused doing so.
k. Entering the cars the seats were found to be all occupied.
l. Clara helped me to work the Problem.
m. Necessity commands me to name myself.

PROSODY.

1. Define Prosody. Def. 1. P. 283.
2. What are Phrases? How many kinds are there? Def. 2.
3. By what is *utterance* modified? Obs. P. 283.
4. Define Rhetorical Pauses. Obs. 2.
5. Are Rhetorical Pauses indicated by marks? Obs. 2.
6. Define Grammatical Pauses. Obs. 3.
7. How many characters are used in punctuation?
8. Name and make them. Obs. 3. P. 283.
9. Describe the office of each of the characters. Obs. 4. P. 283.
10. Give the *Rules, Obs.* and *Exceptions* for the Comma.
11. Define the Semicolon. Give *Rules* and *Obs.* for the Semicolon. R. 8, &c.
12. What is a Colon? Give the Rule for it and *Obs.* R. 9.
13. Define a *Period*, and give the Rule for it. R. 10.
14. What is the office of the Dash? R. 11.
15. Define the mark of Exclamation. R. 12.
16. What is the mark of Interrogation? Rule for its use? R. 13.

GRAMMATICAL AND RHETORICAL SIGNS.

1. Make the Grammatical and Rhetorical signs.
2. For what is the Apostrophe used? Def. 3. P. 289.
3. What are Quotation marks, and when used? Def. 4.
4. What is the Hyphen? How used? Def. 5. P. 289.
5. Describe the Bracket and its use. Def. 6.
6. What is the Parenthesis? Def. 7.
7. Define the References and illustrate their use. Def. 8.
8. What is the Brace used for? Def. 9.
9. Illustrate the use of the Inflections. Def. 10.
10. What do the Measures indicate? Def. 11.
11. When is the Caret used? Def. 12.
12. Explain the use of the Diæresis. Def. 13.
13. When is the Index used? Def. 14.
14. Define the Section. Def. 15.
15. What do you understand by the Paragraph? Def. 16.

16. Define Accent and Emphasis.   Def. 17, 18.
17. What do you understand by Composition?   Def. 19.
18. How many kinds of Composition are there?   Def. 19.
19. What is the difference between Prose and Poetry?   Def. 19.
20. Of what does verse consist?   Def. 20.
21. Describe Lyric Poety, and its various kinds.   Def. 21.
22. Describe Epic Poetry.   Give examples.   Def. 22.
23. What is Dramatic Poetry?   Give examples.   Def. 23.   ♣
24. Define Didactic Poetry.   Def. 24.
25. What is a Charade?   Def. 25.
26. Give the proper name to the following:

> "Swans sing before they die; 'twere no bad thing
> Should certain persons die before they sing."

27. Define an Epitaph.   Def. 27.
28. What is Elegiac Poetry?   Def. 28.
29. Define a Sonnet.   Madrigal.   Def. 29, 30.
30. Define Pastoral Poetry.   Ballad.   Def. 31, 32.

VERSIFICATION.

1. Define Versification.   Blank Verse.   Def. 1, 2.
2. What is Rhyming Verse?   What is a Verse?   Def. 3, 4.
3. Define a Hemistich.   Couplet.   Triplet.   Def. 5, 6, 7.
4. What is a Stanza?   Foot?   Trochee?   Iambus?   Def. 8, 9.
5. Define a Pyrrhic.   Spondee.   Dactyl.   Anapest.   Def. 10.

FIGURES.

1. What are Figures?   For what employed?   Def. 1.
2. What is a Grammatical Figure?   Def. 2.
3. Define a Rhetorical Figure.   Def. 3.
4. Define and describe the use of Aphæresis, Apocope.   Def. 4, 5.
5. Also Paragoge, Synæresis, Prosthesis.   Def. 6, 7, 8.
6. Also Diæresis, Syncope, Tmesis, Ellipsis.   Def. 9, 10, 11, 12.
7. Define Pleonasm.   Syllipsis.   Enallage.   Hyperbaton.   Def. 13, 14, 15, 16.

8. Explain the use of a Simile, Metaphor, Allegory. Def. 17, 18, 19.

9. Also Personification, Irony, Hyperbole, Antithesis. Def. 21, 22, 23, 20.

10. Also Monotony, Synecdoche, Apostrophe. Def. 24, 25, 26.

11. Also Interrogation, Exclamation, Vision, Paralepsis. Def. 27, 28, 29, 30.

12. Also Climax, Anti-climax, Alliteration. Def. 31, 32, 33.

13. What do you understand by Acatalectic?

Ans.—A verse in poetry that is complete.

14. Define Catalectic.

Ans.—A verse lacking one syllable.

15. Define Brachycatalectic.

Ans.—A verse lacking two syllables.

16. Define Hypercatalectic.

Ans.—A verse with one too many syllables.

17. What do you understand by Poetic License?

18. Define Long Meter. Short Meter. Common Meter.

19. What do you understand by Scansion or Scanning?

Ans.—Act of counting the feet in a verse.

20. Define the Cæsura.

Ans.—The Cæsural pause is a metrical division in a verse.

21. Scan the following, and give the Cæsural pause. Tell what kind of measure each verse is, whether the verse is Catalectic, Acatalectic, Brachycatalectic, or Hypercatalectic: P. 297.

"On a mountain stretched beneath a hoary willow." P. 297.
"Rouse him like a rattling peal of thunder." P. 297.
"Lo the sacred herald stands." P. 297.
"Oh I have loved in youth's fair vernal morn." P. 297.
"There is a calm for those who weep." P. 297.
"But we steadfastly gazed on the face of the dead." P. 297.
"Earth has no sorrows that Heaven can not heal." P. 297.
"And there lay the rider distorted and pale." P. 298.
"On the cold cheek of death smiles and roses are blending,
    And beauty immortal awakes from the tomb." P. 299.

ABBREVIATIONS.

1. What are Abbreviations?

2. Define the Abbreviations wherever they occur, and give the

Abbreviations for terms wherever found; and give *Sentences* using the abbreviations and terms correctly.

3. A. A. S.—A. B.—M. A.—A. B.—C. F. M.—A. C.

4. Acct.—Before Christ.—Anno Domini.—Adjt. Gen.

5. Administration.—Act.—A. & F. B. S.—Agent.

6. Give the Abbreviations for every State and Territory, and every month in the year.

7. Ald.—Altitude.—A. M.—American.—Anon.—Apoc.

8. April.—Archb.—Article.—Asst. Sec.—A. S. S. U.—Atty. Gen.

9. B. A.—B.—B. L.—Bls.—Bm.—B. R.—Brig. Gen.—Bro. B. V.

10. Captain. C. C. P.—Chron. County. Colonel.—Com. Arr.

11. Cor. Cr. C. P.—C. P. C. P. S.—C. S.—C. W.—A. C.

12. Dollar. Dwz. D. P.—Dr.—Dv.—Dut. Editor.—E. E. E. I.

13. Epistle. England. Esquire. Etal and so forth. Exec. Com.

14. F. A.—F. E. S.—F. R. S.—G. B.—Gen.—Gov.—H. B. M.

15. H. E. I. C.—H. M. Honorable.—Hon. Men. H. R. H. I. e.

16. I. N. R. I.—Unknown.—I. O. O. F.—I. O. O. G. T.— Isaiah.

17. J. V. D.—K. B.—K. C. B.—K. M.—L. C. Avet. of Song.

18. Noon. M. C.—M. D. Sentence.—M. R. A. S.—M. W. —N. B.

19. New Mexico. N. F.—N. S.—N. F.—Obedient.—O. F. O. U. A.

20. P. Payt.—Ph. A.—P. M. G.—President.—Professor. Prox. P. S.

21. Q. E. D.—Q. C.—Q. R.—R. A. R. E.—Reg. Regt.— Rep. R. M.—R. R.

22. Rt. Hon.—Rt. Rev. R. M.—S. A.—S. A. S.—Schr.— Ser'l.

23. Sen. S. J. C.—S. P. Q. R.—S. S.—S. T. A.—T. O.—Tr. —Trans.

24. Upper Canada. U. E. I. C.—U. K.—Univ. U. S.—U. S. A.

25. U. S. M. Vers. Vol.—V. P.—V. R.—W. I.—W. R. Xt. &c.

CHAPTER VIII.

ARITHMETIC.

REMARK.—It is unnecessary to speak of the importance of a thorough knowledge of Arithmetic. Most candidates are better prepared in this branch than in the others in which they are called upon to be examined. Those who fail are usually deficient in the *reasons* for the principles given. All candidates should therefore clearly understand the *reasons* for every point stated.

NOTE.—The references in the questions on Arithmetic are to "*Davie's University Arithmetic.*" Ex. stands for example. Art. for article.

1. What is Arithmetic? Define number. Unit. Primary base. Art. 4, 3, 1.

2. Define Proposition. Analysis. Operation. Rule. Art. 5, 6, 7, 8.

3. How many fundamental rules? Name them. Art. 9.

4. How many methods of expressing numbers? Name and describe them. Art. 10.

5. What is *Notation?* How many methods? Give them. Art. 12.

6. What are the three principles of this Notation? Art. 13, Note.

7. Describe the *Arabic Notation,* and give the laws governing it. Art. 14.

8. On what does the unit of a figure depend? Art. 19.

9. What is the unit of the place on the right? Art. 19.

10. Define *Numeration,* and give the rule for reading numbers. Art. 20.

11. Give the rule for writing numbers. Art. page 28.

12. What is an abstract number? Concrete number? Simple number? Art. 25.

13. Define quantity. A compound denominate number. Art. 28.

14. When several figures are simply written by the side of each other, what does the language imply? Art. 29.

15. In the English currency how many units of the lowest denomination make one of the second? How many of the 2nd one of the 3rd? Art. 30.

16. In Avoirdupois weight how many of the lowest make one of the second? Art. 31.

17. Define a scale. Tell how many kinds there are. What are they? Art. 32.

18. What is the Scale in English currency? Avoirdupois weight? Art. 32.

19. What is the scale in the common system of numbers? Art. 32.

20. If a row of 0's be written, what does the language of figures determine? Art. 33.

21. What is such a system called? How does the unit change? Art. 33.

22. Are the numbers used in United States money abstract or concrete? Art. 34.

23. According to what scale do the units change? Art. 34.

24. How are dollars reduced to cents? From cents to mills? Art. 34.

25. What is an aliquot part? Name the aliquot parts of a dollar. Art. 34.

26. In English currency is the scale uniform or varying? Art. 35.

27. How many general methods are there of forming numbers from the unit one? What is the first? The second? Art. 37.

28. Into how many general classes may the units of Arithmetic be arranged? Art. 38.

29. Name the different classes.

30. Give the various signs used in Arithmetic. Art. 39.

31. What will be the excess over exact nines in any number expressed by a simple significant figure? Art. 40.

32. How may the excess over exact nines be found in any number whatever? Art. 40.

33. Define *Reduction*. How do you change yards to feet? Inches to feet? Art. 41.

34. Give the rules to reduce numbers from a lower to a higher, and from a higher to a lower order. Art. 41.

35. Define *Addition*. Give rule and reason. Art. 42.

36. What is the proof of an operation in Addition?

37. How many methods of proof are there? *Explain each.* Art. 43.

38. What is the reading process in Addition? Art. 44.

39. Define Subtraction. Art. 45.

40. Define Minuend. Subtrahend. Remainder. Art. 45.

41. Give the rule and reason for Subtraction.

42. How do you prove Subtraction?

43. Explain the process of reading in Subtraction. Art. 47.

44. How do you find the difference of time between two dates? Art. 49.

45. In this computation what part of a year is a month? Art. 48.

46. How many days are reckoned to the month? Art. 48.

47. What time does the civil day begin and end? Art. 48.

48. From one eagle, five dollars, six dimes, and ten cents, take five dollars, seven cents and four mills. Art. 48. Ex. 45.

49. From a stack of hay containing nine tons three qr. 20 lbs. I sold 4 tons 17 cwt. 22 lbs.; how much had I left? Art. 48. Ex. 56.

50. Explain the process and give the reasons for subtracting when the figure in the subtrahend is greater than the same unit in the minuend. Art. 45.

MULTIPLICATION.

1. Define Multiplication. Multiplicand. Multiplier and Product. Art. 49.

2. Why is Multiplication called a short method of Addition? Art. 49. (Note.)

3. How many parts are there in every operation in Multiplication? Art. 50.

4. What are the multiplier and multiplicand called? Art. 49.

5. Tell how many principles follow from the definition of Multiplication. Art. 50.

6. In how many ways may 6×4 be multiplied together? Art. 51.

7. How do the products compare with each other? What does this prove? Art. 51.

8. What is a composite number? Give an example. Art. 52.

9. What are the factors of 9, 8, 12, 16, 11, 19, 108? Art. 52.

10. What is a prime number? Give an example. Art. 53.

11. If several factors be multiplied together will the product be altered by changing their order? How do you multiply by a composite number? Art. 53.

12. What is one factor ending in 0? In two 0's? In three 0's? Art. 53.

13. Explain the process of multiplying by 627. By 214. Art. 54.

14. Explain the five principles which follow from this analysis. Art. 54.

15. What is a partial product? Give the rule for Multiplication. Art. 54.

16. Why do you place the first figure of each product under its own multiplier?

17. What must be observed in the multiplication of U. S. money? Art. 54.

18. How many ways can you prove Multiplication? Art. 55.

19. Give the first, with reasons. Second and third, with reasons. Art. 55.

20. Give the proof by counting out the 9's. Art. 55.

21. Do you consider the method by counting out the 9's a sure test for the accuracy of your product? Art. 56.

22. What do you understand by contractions in Multiplication? How do you multiply when there are 0's in one or both factors? Art. 55.

23. How far would a vessel sail in 9 days of 24 hours each, at the rate of 15 miles an hour? Art. 55. Ex. 50.

24. At the same rate how long would a vessel be in sailing from Kenosha, Wis., to Singapore, India?

25. Give the course of the ship in the above example.

DIVISION.

1. Define Division. Dividend. Divisor. Quotient. Remainder. Art. 57.

2. How many parts in every division? Name them. Art. 58.

3. How many signs in Division? Make them. Art. 58.

4. What is Short Division? Explain the process. Art. 59.

5. Give the rule for division of numbers. Art. 60.

6. How many operations in Long Division? Name them. Art. 60. (Note.)

7. Give the reasons for every step taken in Long Division. Art. 60.

8. When the divisor is greater than the dividend, how will the quotient compare with one? What part will the quotient be of one? Art. 60.

9. How many methods are there for Division? Name them, with the reasons for each. Art. 61.

10. Are you satisfied with the proof by 9's? Art. 61.

11. How long will 9125 loaves of bread last 5 families if each family consume five loaves a day? Art. 61. Ex. 61.

12. If iron is worth 2 cents a pound, how much can be bought for $67? Art. 61.

13. What did you say were contractions in Multiplication? Art. 62.

14. How do you multiply by 25? Art. 63.

15. How do you multiply when the multiplier contains a fraction? Art. 64.

16. How do you multiply by $12\frac{1}{2}$? Art. 65.

17. How do you multiply by $33\frac{1}{3}$? Art. 66.

18. Give the process and reason for multiplying by 125. Art. 66.

19. What are contractions in Division? Art. 68.

20. Give the rules and reasons for dividing by 25. By $12\frac{1}{2}$. By $33\frac{1}{3}$. By 125. Art. 69.

21. Under how many points of view may Division be regarded? Name them. Art. 69.

4

22. What is the rule and reason for Division, when the divisor is a composite number? Art. 70.

23. When there are remainders in Division, how do you find the true remainder in units of the dividend? Art. 70.

24. How do you divide when the divisor is 1 with ciphers annexed? Art. 71.

25. What is the rule and reason for division when the divisor contains significant figures with ciphers annexed? Art. 72.

26. How do you divide when the divisor contains a fraction? Art. 73.

27. What does the analysis of a practical question require? Art. 74.

28. How do you find the cost of any number of things when the price of unity and the number of things are given? Art. 71.

29. How do you find the cost of articles sold by the hundred or thousand? Art. 77.

30. How do you find the cost of articles sold by the ton? Art. 76.

31. What is the object of division abstractly? How many objects has it practically? Name and give the rules and reasons for each. Art. 77.

32. What is Practice in Arithmetic? Give the rules for solving questions by it. Art. 77.

33. What is an Aliquot Part? Give the table of Aliquot Parts of $1. Art. 77.

34. Find by Practice what will be the cost of 335 bushels of potatoes at 3s. 6d.=3½s. a bushel. Art. 77. Ex. 20.

35. Of what number is 365 both a divisor and quotient?

LONGITUDE AND TIME.

1. How is the equator of the earth supposed to be divided? Art. 78.

2. How does the sun appear to move, and what is a day? Art. 79.

3. How far does the sun *appear* to move in one hour? Art. 79.

4. How do you reduce degrees of longitude to time? Art. 80.

5. How do you reduce minutes of longitude to time? Art. 80.

6. What is the hour when the sun is on the meridian? Art. 81.

7. When the sun is on the meridian of any place, how will the time be for all places East? West? Art. 81.

8. If you have the difference of time, how do you find the true time? Art. 81.

9. How do you reduce time to degrees and minutes of longitude? Art. 82.

10. Washington is in longitude 77° 2' west. New Orleans in 89° 2' west. When it is 9 o'clock A. M. at Washington, what is the time at New Orleans? Art. 82. Ex. 5.

11. If the difference of time between London and Oregon city is 8 hours, what is the difference of longitude? Art. 82. Ex. 9.

12. If a man travel 146 miles 7 furlongs 14 rd. 14 ft. in 5 days, how much is that for every day? Give the reasons. Art. 82. Ex. 61.

13. How long will it take to count 20 millions at the rate of 80 per minute? Art. 82. Ex. 11.

14. What time would it be in Chicago when it was 12 M. at Boston? Art. 82. Ex. 95.

15. In 189 mi. 3 fur. 6 rd. 1 ft. how many feet? Art. 82. Ex. 58.

PROPERTIES OF NUMBERS.

1. What is an integral number? Art. 83.

2. When is one number said to be divisible by another? Art. 84.

3. Define a composite number. Prime number. Art. 85, 86.

4. When are two numbers prime to each other? Art. 87.

5. To what product is every number equal? Art. 89.

6. How do you find the prime factors of any number? Art. 89.

7. How do you find the prime factors common to several numbers? Art. 90.

8. What even numbers are prime? What numbers will 2 divide? Art. 91.

9. What numbers will 3 divide? 4? 5? 6? Art. 91.

10. When will the divisor exactly divide the dividend? Art. 91.

11. When will any number divide a product, and why? Art. 91.

12. When will a number divide the sum of 2 numbers? Art. 91.

13. When will it divide either of them separately? Art. 91.

14. When will a number exactly divide the difference of 2 numbers? Art. 91.

15. If a number divide the dividend and divisor, what other number will it always divide? Art. 91.

16. What is a common divisor of 2 or more numbers? Art. 92.

17. What is the greatest common divisor of two or more numbers? How do you find the greatest common divisor of two or more numbers? Give reasons. Art. 93.

18. What is the rule when the numbers are large? Art. 94.

19. What is the greatest common divisor of 4617, 7695, 7642 and 3038? Art. 94. Ex. 6.

20. Three persons, A, B and C, agree to buy a lot of 63 cows, at the same price per head, provided each man can thus invest his whole money. A has $286, B $462, and C $638; how many cows could each man purchase? Art. 94. Ex. 11.

21. Define a multiple of a number. Art. 95.

22. What is a common multiple of two or more numbers? Art. 96.

23. The least common multiple of two or more numbers? Art. 97.

24. How do you find the least common multiple of two or more numbers? Art. 98.

25. Find the least common multiple of 9, 18, 27, 36, 45, 54. Art. 98.

26. Define Cancellation. On what *principle* does it depend? Art. 99.

27. How do you perform the operations of cancellation? Art. 100.

28. What is the quotient of 64 times 840 multiplied by 9 times 122, divided by 32 times 560, multiplied by 4 times 31. Art. 100. Ex. 10.

29. Give reasons for the solution of the last question. Art. 100.

30. What is the quotient of $2 \times 4 \times 8 \times 13 \times 7 \times 16$ divided by $26 \times 14 \times 8$. Art. 100. Ex. 1.

COMMON FRACTIONS.

1. What is a unit? The unit of a fraction? A fractional unit? Art. 101.

2. How do you distinguish between the one and the other? Art. 102.

3. May a fractional unit become the base of a collection? Art. 102.

4. What is a fraction? How are fractions expressed? Art. 102.

5. What is the lower number called? The upper number? What does each denote? Art. 102.

6. What is the primary base of every fraction? Art. 102.

7. How many units have been divided to obtain 6 thirds? Art. 103.

8. To obtain 9 halves? 12 fourths? Art. 103.

9. How may a whole number be expressed fractionally? Art. 104.

10. Does this change the value of the quantity? Art. 104.

11. If the numerator be divided by the denominator what does the quotient show? Remainder show? Art. 105.

12. What form has the fraction? Art. 105.

13. What seven principles can you mention belonging to fractions? Art. 105.

14. If the fraction is one and the fractional unit one 90th, express 9 fractional units. Also 89. Art. 105. Ex. 5.

15. What is a proper fraction? Give examples. Art. 106.

16. What is an improper fraction? Why so called? Art. 107.

17. Define simple fraction. May it be proper or improper? Art. 108.

18. Define a compound fraction. Give examples. Art. 109.

19. What is a mixed number? Give examples. Art. 110.

20. Define complex fractions. Give examples. Art. 111.

21. How many terms has every fraction? Art. 112.

22. How may all the whole numbers be formed? Art. 113.

23. How may the fractional units be found? Art. 114.

24. What part of one is one half? What part of 1 is every fractional unit? Art. 114.

25. What is proved in proposition first? Art. 114.

26. Give proposition II. and the principle involved in it. Art. 115.

27. Give proposition III. and the reason of it. Art. 116.

28. Give proposition IV. and the reason therefor. Art. 117.

29. If both terms of the fraction be multiplied by the same number or quantity what effect will it have on the value of the fraction? Art. 119.

30. If both terms of the fraction be divided by the same quantity how will the value of the fraction be effected? Art. 119.

31. Give reasons for the last three propositions.

REDUCTION OF FRACTIONS.

1. *Define reduction of fractions.* Art. 120.

2. When is a fraction in its lowest terms? Art. 120.

3. How do you reduce a fraction to its lowest terms? Art. 121.

4. How do you reduce an improper fraction to its equivalent whole or mixed number? Art. 122.

5. Reduce $39\frac{7}{8}$ to its equivalent improper fraction, and give rule and reason. Art. 123.

6. Change 19 to a fraction whose denominator shall be 9. Give rule and reason. Art. 123.

7. Reduce $\frac{1}{2} \times \frac{1}{3} \times \frac{1}{4} \times \frac{1}{5}$ to a simple fraction, and give *rule* and *reason.* Art. 124.

8. Reduce $\frac{2}{3}$, $\frac{4}{5} + \frac{3}{4}$, to a common denominator. Give *rule* and *reason.* Art. 125.

9. How may the work often be shortened? Art. 125.

10. How do you find the least common denominator of several fractions? Give *rule* and *reason.*

ADDITION OF FRACTIONS.

1. What is the sum of two or more fractions? Art. 126.

2. Define addition of fractions. How many cases are there? Art. 126.

3. Add $\frac{1}{2}+\frac{3}{4}+\frac{6}{8}$ and $\frac{2}{3}$, and give rule and reason. Art. 127.

4. What is the sum of two fractions equal to when each numerator is equal to one? Art. 128.

5. How do you add fractions having different denominators? Art. 129.

SUBTRACTION OF FRACTIONS.

1. Define subtraction of fractions. Art. 130.

2. How many cases are there? Art. 130.

3. From $\frac{3}{4}$ take $\frac{1}{2}$, and give rule and reason. Art. 131.

4. What is the difference between two fractions whose numerators are each one? Give reason. Art. 132.

5. How do you subtract one mixed number from another? Art. 133.

6. From $\frac{4}{5}$ of a ton take $\frac{5}{8}$ of 12 cwt., and give reason. Art. 134.

MULTIPLICATION OF FRACTIONS.

1. Define multiplication of fractions.

2. Give rule and reason for multiplying one fraction by another. Art. 135.

3. When the multiplier is less than one, what part of the multiplicand is taken? Art. 135.

4. Does multiplication of fractions always imply increase? Art. 135.

5. What part is the product of the multiplicand? Art. 135.

6. What do you do when either factor is a whole number? Art. 135.

7. Multiply $\frac{1}{2}$ of $\frac{7}{8}$ by $\frac{4}{5}$ of $\frac{8}{10}$, and give reasons. Art. 135.

8. Bought a book for $\frac{10}{15}$ of a dollar, and a knife for $\frac{5}{12}$ as much; how much did I pay for the knife? Reason. Art. 135.

9. If I own $\frac{7}{15}$ of a farm and sell $\frac{9}{14}$ of my share, what part of the whole farm do I sell? Why? Art. 135. Ex. 46.

10. A owned $\frac{2}{3}$ of 200 acres of land and sold $\frac{2}{3}$ of his share to B, who sold $\frac{1}{4}$ of what he bought to C; how many acres had each? Give reasons and rules for the operations. Art. 135. Ex. 61.

DIVISION OF FRACTIONS.

1. Define division of fractions. Art. 136.

2. Give rule and reason for division of fractions. Art. 136.

3. What do you do when either the dividend or divisor is a whole number? Art. 136.

4. How do you proceed when either of the fractions is a mixed number or a compound fraction? Art. 136.

5. If the terms of the dividend are exactly divisible by the corresponding terms of the divisor, how do you find the quotient? Art. 136.

6. Divide $\frac{2}{3}$ by $\frac{2}{3}$ and give *clearly* the rule and reason. Art. 137.

7. If $\frac{4}{9}$ of $\frac{2}{3}$ of a barrel of flour will last a family one week, how long will $9\frac{4}{14}$ barrels last them? Why? Art. 137. Ex. 84.

8. Define a complex fraction and give rule and reason for reducing complex fractions to simple ones. Art. 138.

9. What is the sum and difference of $\dfrac{49\frac{5}{8}}{97}$ and $\dfrac{34\frac{3}{8}}{146\frac{3}{14}}$? Art. 138. Ex. 10.

10. A man being asked how many sheep he had, said he had them in three fields; in the first he had 63, which was $\frac{7}{9}$ of what he had in the second, and that $\frac{6}{9}$ of what he had in the second was just 4 times what he had in the third. How many sheep had he in all? Art. 138. Ex. 34.

DUODECIMALS.

1. What are duodecimals? If the unit one foot be divided into 12 equal parts, what is each part called?

2. If one inch be divided into 12 equal parts, what is each part called? Art. 139.

3. What are the indices? Art. 139.

4. By what rules do you operate on duodecimal units? Art. 140.

5. What are the units of the scale? Art. 140.

6. What is multiplication of duodecimals? Art. 141.

7. Give rule and reason for multiplication and division of duodecimals. Art. 142.

8. From a cellar 42 ft. 10 in. long, 12 ft. 6 in. wide, were thrown 158 cu. yds. 17 cu. ft. of earth; how deep was it? Art. 142. Ex. 8.

DECIMAL FRACTIONS.

1. How many kinds of fractions are there? What are they? Art. 143.

2. State the difference between a common and a decimal fraction. Art. 142.

3. When the unit is divided into 10 or 100 equal parts, what is each part called? Art. 144.

4. How are decimal fractions formed? Art. 144.

5. What gives denomination to the fraction? Art. 144.

6. Is the denominator understood? Art. 145.

7. How can you tell what every denominator is?

8. Which way are decimals numerated? Art. 145.

9. On what does the unit of a figure depend? Art. 146.

10. How does the value change from the left toward the right? Art. 146.

11. Give the rule and reason for writing and reading decimals. Art. 146.

12. Write 40 and nine ten millionths. Art. 146. Ex. 8.

13. What is the unit of Federal money? Art. 147.

14. What part of a dollar is a cent? A mill? Art. 147.

15. What effect does annexing a cipher have to the value of a decimal? Art. 148.

16. What effect does prefixing a cipher have to the value of a decimal? Art. 149.

ADDITION OF DECIMALS.

1. What is addition of decimals? Art. 150.

2. What parts of unity may be added together? Art. 150.

3. Give rule and reason for addition of decimals.

SUBTRACTION OF DECIMALS.

1. Give rule and reason for subtraction of decimals.   Art. 151.

2. How many places do you point off in remainder?   Art. 151.

3. From two hundred and twenty-seven thousandths take ninety-seven and one hundred and twenty ten thousandths.   Art. 151. Ex. 24.

MULTIPLICATION OF DECIMALS.

1. After multiplying, how many places do you point off in the product?   Give an example.   Art. 152.

2. When there are not so many places what do you do?   Art. 152.

3. Give rule and reason for multiplication of decimals.   Art. 152.

4. Multiply two hundred and ninety-four millionths by one millionth.   Art. 152.   Ex. 13.

5 What effect does removing the decimal point one place to the right or left have on decimal fractions?   Art. 154.

CONTRACTIONS IN MULTIPLICATION.

1. What is contraction in multiplication of decimals?   Art. 153.

2. What is proposed in the example?   Explain it.   Art. 153.

3. How are the numbers written down for multiplication?   Art. 153.

4. Give the rule and reason for this method.   Art. 153.

5. Where is the first figure of every product to be written, and how do you compensate for the part omitted?   Art. 153.

6. By this method multiply 4745.679 by 751.4549 and reserve only whole numbers in the product.   Art. 153.   Ex. 5.

DIVISION OF DECIMAL FRACTIONS.

1. Define division of decimal fractions.   Art. 155.

2. How do you determine the number of decimal places in the quotient?   Art. 155.

3. Give the rule for the division of decimals.   Art. 155.

4. How do you divide a decimal by 10, 100, or 1000? Art. 156.

5. How many suits of clothes can be made from 34 yds. of cloth, allowing 4.25 yds. for each suit? Art. 156. Ex. 30.

6. If there are more decimal places in the divisor than in the dividend, what do you do? What will the figures of the quotient then be? Art. 157.

7. What do you do after you have brought down all the figures of the dividend? Art. 158.

CONTRACTIONS IN DIVISION.

1. What are contractions in division? Art. 160.

2. Explain the process of making the division. Art. 160.

3. What figures may be omitted in the contracted method? Art. 160.

4. Give the reasons for contractions in division. Art. 160.

5. Divide by this method 98.187437 by 8.4765618. Art. 160. Ex. 4.

REDUCTION OF COMMON AND DECIMAL FRACTIONS.

1. How do you change a common to a *decimal fraction?* Art. 161.

2. How do you change a decimal to the form of a common fraction?

3. A man owns $\frac{7}{8}$ of a ship; he sells $\frac{4}{5}$ of his share: what part is that of the whole, expressed in decimals? Art. 161. Ex. 19.

DENOMINATE DECIMALS.

1. Define a denominate decimal. Art. 163.

2. How do you find the value of a denominate number in decimals of a higher unit? Art. 164.

3. Give rule and reason for finding the value of a decimal in integers of less denominations. Art. 165.

4. What decimal part of a mile is 72 yards? Art. 164. Ex. 29.

CIRCULATING OR REPEATING DECIMALS.

1. How many cases are there of changing a vulgar to a decimal fraction? What are they? Art. 166.

2. What distinguishes one of these cases from another? Art. 166.

3. How can you tell when a vulgar fraction can be exactly expressed decimally? Art. 167.

4. How many decimal places will there be in the quotient? Art. 167.

5. Can ⅓ be exactly expressed decimally? Art. 168.

6. To what does the value of this quotient approach? Art. 168.

7. When does it become equal to one third? Art. 168.

8. Define a circulating decimal. Art. 169.

9. What is a repetend? Give an example. Art. 170.

10. What is a single repetend? A compound repetend? Pure repetend? Mixed repetend? Similar repetend? Art. 171–175.

11. What are dissimilar repetends? Conterminous repetends? Art. 177.

12. What are Similar and Conterminous repetends? Art. 178.

13. Give the *rule* and *reason* for reducing a pure repetend to its equivalent common fraction. Art. 178.

14. How do you find the value of a mixed repetend? Art. 180.

15. How do you add circulating decimals? Art. 183.

16. Give the *rules* and *reasons* for Subtraction, Multiplication, and Division of circulating decimals.

17. Multiply 45.1'3 by '245' and divide 3.'753' by '24'. Art. 186. Ex. 8.

CONTINUED FRACTIONS.

1. What is a continued fraction? Art. 187.

2. Define an approximating fraction. Art. 188.

3. Place ²⁷⁄₄ under the form of a continued fraction and find the value of each approximating fraction. Art. 188. Ex. 5.

RATIO AND PROPORTION.

1. Define ratio. Proportion. Art. 189.

2. From how many terms is a ratio derived? Art. 190.

3. What is the first term called? The second? Which is the standard? Art. 190.

4. How may the ratio of two numbers be expressed and read? Art. 191.

5. What are proportional terms? Art. 192.

6. Which are the extremes of a proportion? The means? Art. 193.

7. What is the product of the extremes equal to? Art. 194.

8. On what principle is the rule for proportion founded? Art. 194.

9. What is Simple ratio? Compound ratio? Art. 195.

10. Define and give the rule of Three, and reason for same. Art. 198.

11. How do you state a question by the rule of Three? Art. 199.

12. At what time between 6 and 7 o'clock will the hour and minute hands of a clock be exactly together? Art. 199. Ex. 46.

13. A can do a piece of work in 3 days; B, in 4 days; C, in 6 days: in what time will they all do it, working together? Art. 199. Ex. 49.

CAUSE AND EFFECT.

1. Define Causes, Simple and Compound. Art. 200.

2. What are Effects, Simple and Compound? Art. 201.

3. What do we infer from the nature of causes and effects? Art. 202.

4. When are two numbers directly proportional? Art. 205.

5. When are two numbers inversely proportional? Art. 205.

6. If two numbers are inversely proportional, what is either equal to? Why? Art. 207.

7. If 72 horses eat a certain quantity of hay in 76 weeks, how many horses will consume the same in 90 weeks? Art. 208. Ex. 28.

COMPOUND PROPORTION.

1. Define Compound Proportion, and tell what it embraces. Art. 209.

2. What is always required? Art. 209.

3. Give the rule and reason for compound proportion. Art. 210.

4. If 5 compositors in 16 days, working 14 hours a day, can compose 20 sheets of 24 pages each, 50 lines in a page, and 40 letters in a line, in how many days, working 7 hours a day, can 10 compositors compose 40 sheets of 16 pages in a sheet, 60 lines in a page, and 50 letters in a line? Art. 210. Ex. 19.

PARTNERSHIP.

1. Define Partnership. Partners. Capital or Stock. Art. 211.

2. What is dividend? Loss? Art. 211.

3. Give *rule* and *reason* for Partnership. Art. 212.

COMPOUND PARTNERSHIP.

1. Define Compound Partnership. Art. 218.

2. Give the REASON, and not the rule, for compound partnership.

3. Where men take an interest in a mining company, A puts in $480 for 6 months, B a sum not named for 12 months, and C $320 for a time not mentioned; when the accounts were settled A received $600 for his stock and profit, B $1200 for his, and C $520 for his; what was B's stock and C's time? Art. 213. Ex. 10.

PER CENTAGE.

1. Define per centage. What is the base? Art. 214.

2. Define per cent. Rate per cent. Art. 215.

3. How do you find the per centage of any number? Art. 216.

4. How do you find the per cent. which one number is of another? Art. 217.

5. How do you find the base when the per centage is added to or subtracted from the base? Art. 218.

6. What per cent. of 800 is eleven? Art. 217. Ex. 9.

7. A grocer purchased a lot of teas and sugar, on which he lost 16 per cent. for selling them for $4200; what did he pay for the goods? Art. 218. Ex. 4.

INTEREST.

1. Define interest. Principal. Amount. Art. 219.

2. Define rate of interest. What does per annum mean? Art. 219.

3. How do you find the interest of any principal for any number of years? Give the analysis, with reasons. Art. 220.

4. How do you find the interest for any time, at any rate? Art. 221.

5. How do you find the interest on any principal for any time, at any rate? Art. 221, 222.

6. Gave a note of $560.40, March 14th, 1855, on interest after 90 days. What interest was due Dec. 1st, 1856, at 10 per cent.? Art. 222. Ex. 27.

7. How do you find the interest when the principal is in pounds, shillings and pence? Art. 223.

8. How many parts are there in every question in interest? Art. 224.

9. How many of these must be known before the remainder can be found? Art. 224.

10. How do you find the *interest* when you know the *principal, rate* and *time?* Art. 224.

11. How do you find the *principal* when you know the *interest, rate* and *time?* Art. 224.

12. How do you apply the formula to *any case?* Art. 224.

13. Give rule and reason for Partial Payments. Art. 225.

14. For value received we jointly and severally promise to pay Jones, Mead & Co. or order, four hundred and fifty dollars on demand, with interest at 8 per cent. Manning & Bros.

The following indorsements were made on this note:

Sept. 25th, 1851, received $85.60. July 10th, 1852, received $20. June 6th, 1853, received $150.45. Dec. 28th, 1854, received $25.12½. May 5th, received $169. What was due Oct. 18th, 1857? Art. 225. Ex. 6.

COMPOUND INTEREST.

1. Define Legal interest. Compound interest. Art. 227.

2. Give rule for computing compound interest. Art. 227.

3. Find by the table what $75 will amount to in 10 years 4 mo. 21 days, at 7 per cent. compound interest. Art. 227. Ex. 13.

4. What will be the compound interest of $200 for 1 year 7 mo. 9 da., at 5 per cent.? Art. 227. Ex. 14.

DISCOUNT.

1. Define Discount. What is the face of a note? Present value? Art. 228.

2. What is the discount on any note? Art. 230.

3. Give the rule for finding the discount. Art. 230.

4. What sum of money will amount to $2500 in 2 years 7 mo. 12 da., at 12 per cent.? Art. 230. Ex. 5.

5. Which is the more advantageous, to buy sugar at $7\frac{1}{2}$ cents a pound on 4 months credit, or at 8 cts. a pound on 6 months, at 6 per cent. interest? Art. 230. Ex. 13.

6. Bought land at $10 an acre; what must I ask per acre if I abate 10 per cent. and still make 20 per cent. on the purchase money? Art. 230. Ex. 14.

BANKING.

1. Define *Bank* and *Banking*. *Bank Bills*. Art. 231.

2. What is a promissory note? Accommodation note? Art. 231.

3. What are BUSINESS NOTES? Days of grace? Art. 231.

4. Write a note payable to bearer. A *joint note*. One payable at a bank. A *negotiable note*. Art. 231.

BANK DISCOUNT.

1. Define Bank Discount. When is the interest paid? Art. 232.

2. How is interest calculated by the customs of banks? Art. 232.

3. What is the bank discount and proceeds of a note of $500 at $6\frac{1}{2}$ per cent., payable in $8\frac{1}{2}$ months? Art. 232. Ex. 3.

4. What is the difference between the true and bank discount of $10,000, payable in $4\frac{1}{2}$ months at 8 per cent.? Art. 232. Ex. 8.

5. How do you find the face of a note of a required present value? Art. 232.

COMMISSION.

1. Define Commission and tell how you find the amount of commission on a given sum. Art. 234.

2. How do you find the amount to be invested exclusive of the commission? Art. 234.

3. A town collector received 4½ per cent. for collecting a tax of $2564.25; what was the amount of his per centage? Art. 234. Ex. 7.

4. A bank fails and has in circulation bills to the amount of $267581; it can pay 9½ per cent.: how much money is there on hand? Art. 234. Ex. 8.

STOCKS AND BROKERAGE.

1. Define *Corporation. Charter. Capital* or *Stock.* Shares. State Stocks. United States Stocks. Art. 236, 237.

2. What is par value? Market value? Art. 238.

3. When is the stock said to be above par? Below par? Art. 238.

4. What is dividend? On what is it estimated? Art. 239.

5. Define Brokerage and tell how you find the value of stock which is above par. Art. 241.

6. The par value of 257 shares of bank stock was $200 a share; what is the present value of all the shares, the stock being at a premium of 15 per cent.? Art. 241. Ex. 3.

7. How do you find the sum which will purchase a given amount of stock at par value? Art. 242.

8. How do you find the rate of interest on an investment when the stock is above or below par? Art. 243.

9. How do you find which is the best investment? Art. 244.

10. Which will yield the largest profit, 8 per cent. stock at a premium of 20 per cent., or 5 per cent. stock at 80 per cent.? Art. 244. Ex. 3.

PROFIT AND LOSS.

1. Define Profit and Loss, and give rule and reason for finding profit or loss. Art. 240.

2. How do you find the cost when you know the per cent. and the amount of sale? Art. 246.

3. How do you find the selling price of an article so as to gain or lose a certain per cent.? Art. 247.

4. How do you find the per centage when you know the gain or loss? Art. 248.

5. Bought a piece of cotton goods for 6 cents a yard, and sold it for 7½ a yard; what was my gain per cent.? Art. 248. Ex. 2.

6. If a merchant sell tea at 66 cents a pound and gain 20 per cent., how much would he gain per cent. if he sold it at 77 cents a pound? Art. 248. Ex. 10.

INSURANCE.

1. Define *Insurance*. *Policy*. *Premium*. Art. 251.

2. How many cases are there which arise in insurance? What are they? Art. 252.

3. Give the rule for finding the premium. Art. 253.

4. What is life insurance? Art. 254.

5. A merchant paid $1920 insurance on his vessel and cargo, which was 5½ per cent. on the amount insured; how much did he insure? Art. 253. Ex. 13.

6. What do you understand by the expectation of life? Art. 255.

7. What may be calculated from the necessary facts? Art. 256.

8. What will be the annual premium for insuring a person's life who is 60 years of age for $2000, at the rate of $4.91 on one hundred dollars? Art. 256. Ex. 5.

ENDOWMENTS.

1. Define Endowments. What does a table of endowments show, and what may be found from the table? Art. 257.

2. What endowment at 21 can be purchased for $961 paid at the age of five years? Art. 257. Ex. 2.

ANNUITY.

1. What is an annuity? Present value of an annuity? Art. 258.

2. How do you find the present value of an annuity for a given rate and time? Art. 258.

3. What is the present value of an annuity of $1500 a year for 30 years, the compound interest being reckoned at 5 per cent.? Art. 258. Ex. 3.

ASSESSING TAXES.

1. What is a tax? Poll tax? How generally collected? Art. 259.

2. What is the first thing to be done in assessing taxes? Art. 260.

3. Explain the process of finding the per cent. of tax to be levied on a dollar. Art. 260.

4. How do you form the assessment table? Art. 261.

5. Give the whole process required in making out a school bill, and tell on what principle founded.

CUSTOM HOUSE BUSINESS.

1. Define a port of Entry. Duty. Custom House. Art. 262.

2. What charges are vessels required to pay? Art. 262.

3. Under whose directions are the revenues of the country? Art. 262.

4. How are duties collected? By whom? Art. 262.

5. Define Specific duty. Advalorem duty. Art. 262.

6. What do the laws of Congress direct in relation to foreign goods? Art. 263.

7. Define gross weight. Net weight. Draft. Tare. Art. 263.

8. What are the different kinds of tare? Art. 263.

9. What will be the duty on 225 bags of coffee, each weighing gross 160 lbs., invoiced at 6 cents a pound, 2 per cent. being the legal rate of tare, and 20 per cent. the duty? Art. 263. Ex. 21.

EQUATION OF PAYMENTS.

1. What is Equation of Payments? Art. 264.

2. How do you find the average time of payment? Art. 264.

3. May the equated time be reckoned from any day? Art. 264. (Note.)

4. A note for $500, dated Nov. 6th, 1856, payable in three

months, was given by E to H. On Dec. 3d, E paid $350; when ought the balance to be paid to balance the account? Art. 264. Ex. 4.

ALLIGATION.

1. Define Alligation, and tell into how many parts it is divided. Art. 265.

2. Define Alligation Medial, and tell how you find the price of the mixture. Art. 266.

3. During the seven days of the week the thermometer stood as follows: 70°, 73°, $73\frac{1}{2}$°, 77°, $80\frac{1}{2}$°, and 81°; what was the average temperature during the week? Ex. 5. Art. 266.

ALLIGATION ALTERNATE.

1. Define Alligation Alternate, and tell how it may be proved. Art. 267

2. How do you find the Proportional Parts? Art. 268.

3. How do you find the Proportional Parts when the quantity of one simple is given? Art. 269.

4. How do you find the Proportional Parts when the quantity and the mixture is given? Art. 270.

5. A farmer sold a number of colts at $50 each, oxen at $40, cows at $25, calves at $10, and realized an average price of $30 per head; what was the smallest number he could sell of each? Ex. 3.

6. A merchant has four pieces of calico, each worth 24, 22, 20, 15 cents a yard; how much must he cut from each piece to exchange for 42 yds. of another piece worth 17 cents a yard? Art. 270. Ex. 7.

COINS, CURRENCY AND EXCHANGE.

1. Define *Coins. Currency* and *Exchange.* Art. 271.

2. What is provided by the Constitution of the U. States? Art. 271.

3. How many values may a coin be said to have? Art. 272.

4. Define each value. What is a Bill of Exchange? Art. 274.

5. How many Parties are there to a bill of exchange? Name them. Art. 274.

6. Describe how bills of exchange aid commerce, and name all the Parties to the bill in this example. Art. 275.

7. Define an inland bill. A foreign bill. Art. 276.

8. How is the time determined at which it is made payable? How are bills always drawn? Art. 277.

9. How many bills are generally drawn for the same amount?

10. What is an indorsement in blank? A special indorsement? Art. 280.

11. What is the Person making the bill called? Art. 280.

12. What is the effect of an indorsement? How may a bill drawn to bearer be transferred? Art. 280.

13. What is acceptance? How made? Art. 281.

14. Tell all you can about *making, drawing* and *protesting* bills of exchange, and Par of exchange. Course of exchange. Art. 282.

15. What is the exchange value of a pound Sterling? Art. 287.

16. In what currency are the exchanges between this country and England made? What is the commercial value of the Pound sterling? Art. 288.

17. I have $947.86 and wish to remit to London £364 18s. 8d., exchange being at 8¼ per cent.; how much additional money will be necessary? Art. 289. Ex. 5.

18. Describe the method of exchange with France. Hamburg.

19. What amount in dollars and cents will produce a bill of exchange on Hamburg for 18649 Mares banco, exchange being at 2 per cent. premium? Art. 291. Ex. 1.

ARBITRATION OF EXCHANGE.

1. Define arbitration of exchange. Art. 294.

2. What principle is involved in arbitration of exchange? Art. 294.

3. What is the chain rule? Explain it. Illustrate by an example.

GENERAL AVERAGE.

1. Define Average. General average. Art. 295.

2. How many kinds of average are there? Name them. Art. 296.

3. Under what circumstances will a general average occur? Art. 296.

4. How is the freight valued? Cargo? Ship? Art. 297.

5. How much is charged on account of the Seaman's wages? Art. 297.

6. Explain the Principle by an example. Art. 297. Ex. 1.

TONNAGE OF VESSELS.

1. What is the tonnage of a vessel? Art. 298.

2. To what are coasters subject? Art. 298.

3. What is the government rule for finding the tonnage of vessels? Art. 299.

4. What is the difference between the government rule and the ship-builder's rule? Art. 299.

5. What is the government tonnage of a double-decker, the length being 103 ft., breadth 25 feet 6 inches? Art. 299. Ex. 4.

INVOLUTION.

1. Define Involution. Power. Root of Power. Third Power. Art. 300.

2. What is the exponent of a Power? How written? Art. 301.

3. How many things are connected with every Power? Art. 301.

4. How do you find the Power of a number? Art. 301.

5. Find the cube of $14\frac{2}{3}$. The value $(3.205)^2$. Art. 301. Ex. 36–42.

EVOLUTION.

1. Define Evolution. Square root. Cube root. Art. 302.

2. How do you denote the square root? Cube root? Art. 302.

3. What is a perfect square, and how many are there between 1 and 100? Art. 303.

4. Into how many parts may every number be decomposed? When so decomposed, to what is its square equal? Art. 304.

5. What is the first step in extracting the square root? Art. 305.

6. Give the *rule* and *reason* for extracting the square root. Art. 305.

7. How do you extract the square root of decimal fractions? Art. 306.

8. How of a common fraction? Art. 306.

9. Define a right angle. A Triangle. Art. 307.

10. Define a right-angled triangle. Hypothenuse. Art. 308.

11. In a right-angled triangle to what is square of the Hypothenuse equal? Why? Art. 309.

12. How do you find the Hypothenuse when you know the base and perpendicular? Art. 310.

13. When you know the Hypothenuse and one side, how do you find the other side? Art. 311

14. Find the square root of 225. of $\frac{7}{8}$. Art. 306. Ex. 3 and 13.

15. What length of a rope must be attached to a halter 4 feet long, that a horse may feed over $2\frac{1}{2}$ acres of ground? Art. 311. Ex. 16.

16. Three men bought a grindstone which was four feet in diameter; how much must each grind off to use up his share of the stone? Art. 311. Ex. 17

CUBE ROOT.

1. What is the cube root of a number? Art. 312.

2. When is a number a perfect cube? Art. 312.

3. How many perfect cubes are there between 1 and 1000? Art. 312.

4. Of how many parts is the cube of a number composed? Art. 313.

5. Name and describe them. Art. 313.

6. Give and demonstrate the rule for extracting the cube root of a number. Art. 314.

7. How do you extract the cube root of a common or a decimal fraction? Art. 315.

8. How many places will there be in the root? Art. 315.

9. What is the cube root of 46.656? Of 8.343? Art. 314, 315. Ex. 3, 1.

10. What is the difference between half a cubic yard and a cube whose edge is half a yard? Art. 316. Ex. 6.

11. If I put 2 tons of hay in a stack 10 feet high, how high must a similar stack be to contain 16 tons? Art. 316. Ex. 15.

12. Four women bought a ball of yarn 6 inches in diameter and agreed that each should take her share separately from the surface of the ball; how much of the diameter must each wind off? Art. 316. Ex. 16.

ARITHMETICAL PROGRESSION.

1. Define Arithmetical Progression. Common Difference. Art. 317.

2. What is a decreasing, and what is an increasing series? Art. 318.

3. Which are the means, and which the extremes, of a progression? Art. 318.

4. How many parts are there in every Arithmetical Progression? Art. 319.

5. How many parts must be given before the remaining ones can be found? Art. 319.

6. When you know the first term, the common difference and the number of terms, how do you find the last term? Art. 320.

7. What will $200 amount to in 15 years, at 7 per cent. simple interest; the first year it increases $14, the second $28, and so on? Art. 320. Ex. 3.

8. When you know the extremes and number of terms, how do you find the common difference? Art. 321.

9. How do you find the sum of the series? Art. 322.

10. Having given the first and last terms, and the common difference, how do you find the number of terms? Art. 323.

11. A person proposes to make a journey and travel 15 miles the first day, and 33 miles the last, with a daily increase of $1\frac{1}{2}$; in how many days did he make the journey, and what was the whole distance travelled? Art. 323. Ex. 2.

GEOMETRICAL PROGRESSION.

1. Define Geometrical Progression. Ratio. Art. 322.

2. What is an increasing Series? Decreasing Series? Art. 325.

3. Define the *terms,* means and extremes of a Progression. Art. 326.

4. How many parts are there in every Geometrical Progression? Art. 326.

5. How many must be known before the rest can be found? Art. 326.

6. Knowing the first term, the ratio, the number of terms, how do you find the last term? Art. 327.

7. The first term of a decreasing geometrical series is 729, the ratio ⅓; what is the 10 term? ▪Art. 327. Ex. 3.

8. Knowing the two extremes and the ratio, how do you find the sum of the terms? Art. 328.

9. A merchant engaging in business trebled his capital once in 4 years; if he commenced with $2000, what will his capital amount to at the end of the 12th year? Art. 327. Ex. 6.

10. A laborer agreed to thresh 64 days for a farmer, on the condition that he should give him 1 grain of wheat for the first day's labor, 2 grains for the second, and double each succeeding day; what number of bushels would he receive, supposing a pint to contain 7,680 grains, and what number of ships, each carrying 1000 tons burden, might be loaded, allowing 40 bushels to a ton? Art. 328. Ex. 5.

ANALYSIS.

1. Define Analysis and tell wherein it differs from the *"Rule of Three."*

2. By analysis find the cost of 12½ lbs. of tea at 6s. and 8d. a pound, Pennsylvania currency. Page 329. Ex. 7.

3. A general arranging his army in the form of a square, finds that he has 44 remaining; but by increasing each side by another man, he wants 49 to fill up the square; how many men had he? Page 348. Ex. 103.

4. If a ball 2 inches in diameter weighs 5 pounds, what will be the diameter of another ball of the same material that weighs 78,125 pounds? Page 350. Ex. 120.

5

MENSURATION.

1. Define *Mensuration.* *Surface.* *Square.* Art. 329, 330.

2. What is a *triangle?* *Base* of a triangle? *Altitude?*

3. Which side is the hypothenuse of a right-angled triangle? Art. 331.

4. What is the area of a triangle equal to? What is a rectangle?

5. Define a Parallelogram. Trapezoid. Art. 335.

6. How do you find the area of a Parallelogram? Square? Rectangle, or Trapezoid? Art. 336.

7. What is the area of a trapezoid whose parallel sides are 15 chains and 245 chains, and the perpendicular height 30.80 chains? Art. 337. Ex. 5.

8. Define a *Circle.* *Radius.* *Center.* Art. 337.

9. How do you find the diameter when the circumference is known? Art. 338.

10. What is the area of a circle whose diameter is 5? Art. 339. Ex. 2.

11. How do you find the surface of a sphere? Contents of a sphere? Art. 343.

12. Required the area and contents of the earth, its mean diameter being 7918.7 miles. Art. 343. Ex. 5.

13. How do you find the convex surface of a Prism? Its contents? Art. 346.

14. What is a cylinder? How do you find its convex surface? Art. 348.

15. What are the contents of a cylinder the diameter of whose base is 25 feet, and altitude 15? Art. 349. Ex. 5.

16. Define a pyramid. How do you find the contents of a pyramid? Art. 351.

17. A Pyramid with a square base, of which each side is 15, has an altitude of 24; what are its contents? Art. 351. Ex. 7.

18. Define a cone. How do you find the contents of a cone? Art. 353.

19. What are the contents of a cone whose altitude is 27 feet, and the diameter of the base 20 feet? Art. 353. Ex. 4.

GAUGING.

1. What is a cask gauging? How many varieties of casks are there?

2. Give the rule for finding the mean diameter. Art. 356.

3. How do you find the contents in cubic inches? Art. 357.

4. How many wine gallons in a cask of which the head diameter is 24 inches, bung diameter 36 inches, and length 3 feet 6 inches, the cask being of the second variety? Art. 357. Ex. 4.

MECHANICAL POWERS.

1. How many simple machines are there? Art. 358.

2. Name and describe each. Describe each variety of levers. Art. 361.

3. When is an equilibrium produced in all the levers?

4. What is the proportion between the weight and power? Art. 362.

5. In a lever of the third, the distance from the fulcrum to the weight is 12 feet, and to the power 8 feet; what power will be necessary to sustain a weight of 100 lbs.? Art. 362. Ex. 8.

PULLEY.

1. Define a pulley. How many kinds are there? Art. 365.

2. Does a fixed or movable pulley give any increase of power? Art. 366.

3. What advantage will be gained by several movable pulleys? Art. 367.

4. In two movable pulleys with 4 cords, what power will support a weight of 100 lbs.? Art. 368. Ex. 3.

5. Define an inclined *Plane*. *Wedge*. What used for. Art. 381.

6. Define a *Screw*. Nut. What is the power of a screw? Art. 381.

7. If a power of 300 lbs. applied at the end of a lever 15 feet long will sustain a weight of 282,744 lbs., what is the distance between the threads of the screw? Art. 381. Ex. 4.

UNIFORM MOTION.

1. Define uniform motion. Velocity of a moving body. Art. 383.

2. To what is the space passed over in a unit of time equal? Art. 384.

3. To what is the space passed over in uniform motion equal?

LAWS OF FALLING BODIES.

1. How does the velocity of a falling body change? Art. 386.

2. State and explain the four principles involved in falling bodies. Art. 386.

3. How far will a body ascend when projected upwards? Art. 387.

4. Are the above laws perfectly or only approximately true? Art. 388.

5. A stone is dropped from the top of a bridge and strikes the water in 2.5 seconds; what is the height of the bridge? Art. 388. Ex. 9.

6. A rocket is projected vertically upwards with a velocity of 386 feet; after what time will it begin to fall, and to what height will it rise? Art. 388. Ex. 15.

SPECIFIC GRAVITY.

1. Define specific gravity. What is the standard for measuring the specific gravity of a body? Art. 389.

2. How do you find the specific gravity of a body? Art. 389.

3. A piece of copper weighs 93 grains in air, and 82¼ grains in water; what is its specific gravity? Art. 389. Ex. 1.

4. What weight of mercury will a conical vase contain of which the radius of the base is 9 inches, and the altitude 34 inches, the specific gravity of the mercury being 13.596? Art. 389. Ex. 15.

5. To what is the volume of a vapor or gas proportional? Art. 390.

6. To what is its density proportional?

7. Eight quarts of hydrogen gas are contained in a vessel and submitted to a pressure of 22 lbs.; how many quarts of gas will

there be if the pressure is changed $9\frac{1}{2}$ pounds? Art. 390. Ex. 6.

APPENDIX.

NOTE.—The design and limit of this work require that the questions on this part of *Arithmetic* be comprehensive.

1. Name and tell how many kinds of units there are in Arithmetic. Art. 991.

2. Describe an *abstract unit,* and each unit in its order. Art. 392.

3. Describe the *unit* of *currency. Length. Weight. Surface. Time.*

4. Repeat accurately the *tables* of the *various units* in their *order:* First, U. S. money. Art. 404. 2d, English money. Art. 406.

5. 5th, Table of Linear Measure. Art. 407

6. Cloth Measure. Art. 410.

7. Square Measure. Art. 411. Surveyor's Measure. Art. 412.

8. Cubic Measure. Art. 413. Wine Measure. Art. 414. Beer Measure.

9. Dry Measure. Art. 416. Avoirdupois Weight. Art. 417. Troy Weight.

10. Apothecaries' Weight. Art. 419. Measure of Time. Circular Measure.

11. Miscellaneous Table. *Books* and *Paper.* Art. 422.

REMARK.—Many additional questions might be proposed in this branch. But the *candidate* who answers accurately the foregoing questions, assigning *reasons* for his views, need not fear an Examination before any Board of Examiners in this branch.

Solve the following:

For value received, seven years from date, I promise to pay the Kenosha and Mississippi Cotton Growing Association $7897.86, in *seven equal annual payments,* at seven per cent. compound interest. What *sum must be paid each year?*

KENOSHA, Wis., Jan. 23, 1864. I. S.

CHAPTER IX.

HISTORY.

No Student, much less a Statesman, doubts that a clear and accurate knowledge of History is of intrinsic value in itself; grand in its consequences on nations and men, and the destinies of each. But never so important at any time or to any class, as at this hour, and to the youth of our land. It is not asking then too much, to demand that all Teachers should be *well read* in *General History*.

Note.—The following references in the questions on History are to "Williard's Universal History." P. stands for page.

1. What is History?

Ans.—History is a narrative of past events of individuals, States and Nations.

2. State what you can concerning the *earliest history of man*. P. 33.

3. What can you say of the ancient empire of Assyria? Egypt? P. 36.

4. Mention briefly the history of the Israelites or Jews. P. 39.

5. What can you say of Palestine? Phœnicia? Greece? Troy? P. 43–47.

6. What can you say of the Persian Empire under Cyrus? P. 65.

7. State briefly what you can of Macedonia and Alexander the Great. P. 85.

8. Give a brief sketch of Roman History. P. 88.

9. What became of the empire of Alexander after his death? P. 99.

10. What caused the decline of the Roman Empire? P. 125–140.

11. Mention a few facts connected with the rise of Christianity. P. 144.

12. Describe the nations formed on the ruins of the Roman Empire. P. 184.

13. Describe briefly Mahomet, his flight and religion. P. 191.

14. Give a short narrative of Charlemagne and his efforts. P. 203.

15. State the most prominent events of Britain, Germany and France. P. 207, 215, 220.

16. What can you say of Pilgrimages? Chivalry? The Crusaders? P. 225.

17. Give a brief account of the Greek Empire. Germany. Turkey. P. 231–244.

18. Mention the important events of England. The war of the Roses. P. 251–264.

19. What can you say of Spanish Inquisitions? Italy? P. 272–280.

20. In what war was Spain engaged in the year Columbus discovered America? P. 279.

MODERN HISTORY.

1. At what period does modern history begin? P. 291.

2. Give a brief account of Columbus and the discovery of America. P. 301.

3. What can you say of Martin Luther and the Reformation? P. 315.

4. Mention what you can concerning Henry VIII. John Knox. Cromwell. P. 322.

5. Give a short account of Queen Elizabeth and the events under her reign. P. 329.

6. Give some account of the Huguenots, and the Massacre of Bartholomew. P. 341.

7. State what you can of the Scandinavian nation. Of Gustavus Adolphus. P. 349.

8. Give some account of Henry IV. of France. Of Richelieu. P. 355.

9. Describe the Gunpowder plot. Long Parliament. John Hampden. P. 357.

10. Give some account of the beheading of Charles I. and Oliver Cromwell. P. 363.

11. Give some account of Europe during the war of Spanish succession. P. 385.

12. State the result of the thirteen years' war of the Spanish succession. P. 386.

13. What can you say of Peter the Great?

GENERAL QUESTIONS.

14. Give some account of the treaty of Utrecht and the young Pretender. P. 403, 411.

15. Relate briefly the life and character of " *The Great Commoner.*"

16. State what you can concerning Alfred the Great. P. 209. William the Conqueror. P. 213. Frederick the Great. P. 415.

17. Give an account of Napoleon, his war, and generals. P. 443.

18. Relate the principal events of " *The hundred days.*" P. 463.

19. Describe the " *Holy alliance*" of 1815, and its effect. P. 483.

20. What led to the Partition of Poland? P. 417.

UNITED STATES HISTORY.

1. When and by whom was America discovered?

2. When and by whom was St. Lawrence discovered? P. 364.

3. When and by whom was the first settlement in the U. S. made? P. 364.

4. Who discovered Florida? Where was the first English settlement made in the U. S.? P. 365.

5. What is said of Captain Smith? Relate the heroism of Pocahontas. P. 365.

6. Relate the events and results of the sailing of the May Flower. P. 365.

7. Who came over in the May Flower? Where did they settle? P. 365.

8. What caused the Pilgrims to make new homes in this western wilderness? P. 365.

9. Who discovered the Hudson River? What towns did the Dutch form? P. 366.

10. Give a brief account of the "Old French War." P. 413.

11. Give an account of the battle on the Heights of Abraham. Of Wolf. P. 415.

12. Who was Governor of Virginia in 1753? Who of Canada? P. 423.

13. Whom did the Governor send on a mission to Canada in winter? P. 423.

14. What was the occasion of the Congress of delegates at Albany, in 1754? P. 423.

15. What plan was there drawn up, and by whom? P. 423.

16. What *principles* early found a home in America? P. 424.

17. Was the mother country satisfied with these principles? P. 424.

18. What occurred at Braddock's field? At West Edward? P. 425.

19. Relate the events of the campaign of 1759, under Gen. Amherst and Wolfe. P. 426.

20. Give a clear account of the Stamp Act. P. 426.

21. How did the Americans regard it? Where did their Congress first meet? P. 427.

22. Describe clearly the *occasion* and place of the *first battle*, and its result. P. 427.

23. What can you say of the Continental Congress at Philadelphia? P. 427.

24. When was Washington appointed Commander-in-Chief of the American forces? P. 427.

25. What took place at Boston, on the 17th of March? P. 428.

AMERICAN INDEPENDENCE.

1. What is the Birth-day of the American Independence? P. 431.

2. What was the most disastrous defeat of the war? P. 431.

3. What did Washington do on the 26th of Dec., 1776? P. 432.

4. What *noble* foreigner arrived here in the winter of 1776–77? P. 432.

5. Relate the successes of the British in Pennsylvania. P. 432.

6. Relate the battles of Bennington. Stillwater. Saratoga. P. 432.

7. What important event followed the battle of Saratoga? P. 432.

8. Give an account of the battle of Monmouth and its results. P. 433.

9. Relate the capture of Charleston, the battle of Camden and Eutaw Springs. P. 433.

10. What did Lord Cornwallis threaten to do to " *The boy*" *Lafayette?* P. 433.

11. Relate the operations of Washington until he arrived at Yorktown. P. 434.

12. By whom were the British invested by sea? P. 434.

13. What was the glorious results of these combined operations? P. 434.

14. What other disasters did Great Britain meet? P. 434.

15. By what treaty did Great Britain acknowledge the American Independence? P. 434.

16. What territories did Great Britain lose by this treaty? P. 434.

17. Give an account of the deportment of Washington after peace. P. 435.

18. What can you say of the Articles of Confederation? P. 435.

19. When was the *Constitution* of the U. S. adopted? P. 435.

20. Who was the first President, and who formed his Cabinet? P. 435.

21. Give the history of the members of this first Cabinet during their private lives. P. 435.

22. Give an account of all the Presidents and their Cabinets, in their order.

23. What led to the war of 1812? P. 465.

24. Who was King of England at the time of the American Revolution? P. 465.

25. Who was the Prime Minister of England at the same time? P. 465.

26. What important engagement at New Orleans in 1815? P. 487.

27. When was the treaty of peace signed that ended this war? P. 488.

28. Who was *leader* of the *Nullification Party* of S. C. in 1832? P. 490.

29. What was the result of this effort of S. C.? P. 490.

30. Give the history of the National Bank and its opposition. P. 491.

MEXICAN WAR.

NOTE.—The references in the following questions are to "Willard's Last Leaves." •

1. Who discovered Texas? ANS.—La Salle. Who was he?

2. What led to the Texan Independence? P. 27.

3. Give an account of the Massacre at Goliad. P. 27.

4. What led to the war with Mexico? P. 30 and 31.

5. Give a general account of the Commanders and forces on both sides, and the *various battles* and results of the war. P. 31–105.

SECESSION AND ITS CONSEQUENT REBELLION.

1. Give a brief account of the causes of the Southern Rebellion.

2. When did S. C. Secede? Mississippi? Florida? Alabama? Georgia? Louisiana? Texas? Tennessee? Virginia?

3. State what you can concerning the attack upon and evacuation of Fort Sumter. •

4. When did the President call for 75,000 Volunteers? April 15th, 1861.

5. State what occurred on April 19th, 1861, as the 6th Mass. Reg't were passing through Baltimore.

6. When did the rebels seize the U. S. forts?

7. Relate the events of the first battle of Bull Run, July 21st, 1861.

8. When and by whom was Fort *Hatteras* captured? *New Orleans?*

9. At what engagement was *Gen. Lyon* killed?,

10. State the events incident to the *capture* of *Mason* and *Slidell.*

11. Describe the engagement between the MONITOR and MERRIMAC. Other Naval engagements.

12. Give an account of the battle of *Pea Ridge*, Ark. *Fort Henry*, Tenn.

13. State the events of the *battle* and *surrender* of Fort Donelson to Gen. Grant.

14. When was *Nashville*, Tenn., *occupied* by our forces?

15. Give an account of the occupation of Columbus, Ky., by our forces.

16. What can you say of the *attack* on Island No. 10? Battle of Winchester?

17. Relate the events of the siege of Yorktown. Capture of Fredericksburg.

18. The fight at Strasburg, Va. The battle of Pittsburg Landing.

19. State the events of the Siege of Vicksburg by Gen. Grant, and Port Hudson by Gen. Banks.

20. What caused and followed the evacuation of Corinth by the rebels?

21. Relate the events of the seven days battle on the Peninsula.

22. Also the battles of Fair Oaks, Seven Pines, and White Oak Swamp.

23. Give an account of the battles of Perrysville, Stone River, Chickamauga, *Mission Bridge* and Knoxville.

24. When did the Bill to abolish Slavery in the District of Columbia pass Congress?

25. When did President Lincoln issue his Emancipation Proclamation? Mention any other important events and results of this wicked Rebellion.

NOTE.—The questions on General History could have been multiplied to almost any extent; but the above are sufficient, if the candidates are prepared to answer them; if not, they *are surely* sufficient.

CHAPTER X.
PHYSIOLOGY.

The following references in the questions on Physiology are to CUTTER'S ANATOMY, PHYSIOLOGY AND HYGIENE.

NOTE.—If Agesilaus gave a correct reply when he was asked "What should boys learn?" by saying, "Those things which they will *practice* when they become men:" then it is clear that *teachers* should be *qualified* to *teach Physiology* thoroughly.

ANATOMY.

1. What is Anatomy? How is it divided? P. 13.
2. What is Physiology, and how is it divided? P. 13.
3. What is Vegetable Physiology? Animal Physiology? P. 13.
4. What is Comparative Physiology? What is Hygiene? P. 13.
5. What is the difference between an Organic and Inorganic body? P. 14.
6. How do plants grow? How do animals grow? P. 15.
7. What can you say of disease?
8. Is the study of Physiology important to all? Why?
9. Why is it especially important to Students?

STRUCTURE OF MAN.

10. What can you say of the structure of Man? P. 17.
11. What substances enter into the structure of the human body? P. 17.
12. Define *Fibre. Muscle. Tissue. Organ.*
13. What is the Serous Tissue? Dermoid Tissue? P. 20.
14. What is the Adipose Tissue? Cartilaginous Tissue? P. 22.
15. Define the Osseous Tissue? Muscular Tissue? P. 23.
16. Define the Mucous Tissue. Nervous Tissue. P. 24.
17. What can you say of the Chemistry of the Human body? P. 25.

CHEMISTRY OF THE HUMAN BODY.

18. What is an ultimate element?

19. Name such elements as enter into the composition of Man. P. 25.

20. What is Mucus? *Fibrin?* Gelatin? Albumen? P. 27.

21. What are Bones? Give the anatomy of the bones. P. 29.

THE BONES.

22. What is a natural skeleton? Composition of bones? P. 29.

23. When does true Ossification commence? P. 30.

24. How many bones in the human body? P. 32.

25. How are they divided? Give the anatomy of the bones of the head. P. 32.

26. What are Sutures and their uses? How many bones has the ear? P. 34.

27. How many bones in the trunk? Name them. P. 34.

28. How many bones in the face? Name them. P. 34.

29. Describe the Thorax. Describe the Spinal Column. P. 36.

30. Give the structure of the Vertebræ. P. 36.

31. Give the anatomy and structure of the bones of the upper and lower extremities. P. 39.

PHYSIOLOGY OF THE BONES.

32. Give the Physiology of the bones. P. 48.

33. To what may the bones be compared? P. 48.

34. Give the Hygiene of the bones. P. 53.

35. What effect has exercise upon the bones? P. 53.

36. What effect has inaction on the bones? P. 53.

37. What can you say in regard to teachers requiring their young pupils to remain in one position for a long time? P. 54.

38. How should benches and chairs be constructed in the school room? P. 55.

39. Why should compression of the chest be avoided? P. 56.

40. What should be the position of pupils in the school room? P. 58.

41. What is one cause of rickets? P. 62.

THE MUSCLES.

42. What is a Muscle? Fasciculi? P. 64.
43. Give their Anatomy and their Structure. P. 64.
44. Describe the Diaphragm. To what is it compared? P. 72.

PHYSIOLOGY OF THE MUSCLES.

45. Give the Physiology of the Muscles. P. 76.
46. Give the Hygiene of the Muscles. P. 85.
47. Why do muscles increase in size when exercised? P. 85.
48. Why should not small children be confined in one position for a long time? P. 87.
49. Why should not severe labor be imposed on small children? P. 88.
50. How should exercise be taken? P. 91.
51. What kinds of exercise are best? P. 92.
52. What effect has the mind on the muscular system? P. 93.
53. How should the child be taught to sit at the desk? P. 99.
54. Why have so many pupils failed in acquiring good penmanship? P. 103.

THE TEETH.

55. What can you say of the teeth? P. 105.
56. Where and how are the teeth formed? P. 107.
57. Give the names of the permanent teeth. P. 107.
58. Into how many parts are the teeth divided? P. 108.
59. Give the Physiology of the teeth. P. 109.
60. Give the Hygiene of the teeth. P. 110.
61. Why is smoking injurious to the teeth? P. 111.

THE DIGESTIVE ORGANS.

1. Which are the Digestive Organs? P. 113.
2. Give the Anatomy of the Digestive Organs. P. 113.
3. How many Glands about the mouth? Name them. P. 114.
4. Explain the office of the Stomach. Liver. P. 122.
5. Give the Physiology of the Digestive Organs. P. 124.
6. What is necessary before food can nourish the body? P. 124.

7. Give the Hygiene of the Digestive Organs. P. 129.

8. How much food should be eaten? P. 133.

9. What kinds of food should be eaten in a hot climate? Cold climate?

10. Does the mind have any influence on the Digestive Organs? How and why?

CIRCULATORY ORGANS.

1. Give the anatomy of the Circulatory Organs. P. 154.

2. Describe the heart and its office. P. 155.

3. What are Arteries? Describe the Pulmonary Artery. P. 158.

4. Describe the Aorta, and give its office. P. 159.

5. Describe the Veins, and give their office. P. 160.

6. Give the Physiology of the Circulatory Organs. P. 164.

7. Give the Hygiene of the Circulatory Organs. P. 172.

8. What is the treatment of wounds caused by the bite of rabid animals? P. 179.

9. Define Absorption and Lymphatic vessels. P. 181.

10. Give the anatomy of the Lymphatic vessels. P. 181.

11. Give the Physiology and Hygiene of the Lymphatic vessels. P. 183 and 188.

12. Define Secretion. Exhalants. P. 192.

13. Give the Anatomy, Physiology and Hygiene of the Secretory Organs. P. 192, 193 and 197.

14. Define Nutrition. P. 200.

15. What is the function of the Nutrient vessels. P. 200.

16. Give the Hygiene of Nutrition. P. 205.

RESPIRATORY ORGANS.

1. Give the Anatomy of the Respiratory Organs. P. 209.

2. Name the Respiratory Organs. P. 209.

3. What other organs aid these? P. 209.

4. Describe the Lungs. By what are they enclosed? P. 211.

5. Describe the Bronchia. Trachea. P. 212.

6. Give the Physiology of the Respiratory Organs. P. 217.

7. What is the object of Respiration? Give an experiment

showing that Oxygen changes dark colored blood to a bright red. P. 226.

8. Give the Hygiene of the Respiratory Organs. P. 228.

9. How is the purity of the air affected by Respiration? P. 228.

10. What is said respecting the weight of the blood? P. 228.

11. Why should a School-Room, and all public rooms and sleeping rooms be *well* ventilated? P. 233.

12. How can the size of the chest be diminished? P. 239.

13. Give your opinion about the styles of dress.

ANIMAL HEAT.

1. What is Animal Heat? What is the temperature of the human body? P. 252.

2. Give the Hygiene of Animal Heat. · P. 261.

3. Does age affect the degree of Animal Heat? P. 265.

VOICE.

4. Define voice. Give the Anatomy of the Vocal Organs. P. 268.

5. Give the Physiology of the Vocal Organs. P. 272.

6. Give the Hygiene of the Vocal Organs. P. 274.

7. How should public speakers dress their necks? P. 276.

8. Should students practice Physical exercise? Why? P. 279.

9. Is repetition essential to distinct Articulation? P. 280.

10. How can stammering be remedied? P. 281.

SKIN.

1. What is the skin? Give the Anatomy of the skin. P. 282.

2. Describe fully the skin in all its parts and offices. P. 282.

3. Give the Physiology and Hygiene of the skin. P. 293 and 301.

4. What is the best material for clothing for the different seasons of the year? P. 303.

5. Is bathing beneficial? Why? P. 315.

6. Describe the appendages of the skin. P. 322.

NERVOUS SYSTEM.

1. What is the Nervous System? P. 328.

2. Give the Anatomy of the brain and Cranial nerves. P. 328.

3. Describe the Dura Mater. Pia Mater. Cranial nerves. P. 335.

4. Give the Anatomy of the Spinal Chord. P. 340.

5. Give the Physiology and Hygiene of the nervous system. P. 346 and 358.

6. May too much mental labor be required of students? P. 364.

7. What error prevails in the present system of education? P. 366.

8. What persons require the most sleep? P. 369.

9. Describe sensation and the *sense* of *touch*. P. 378.

10. Give the Hygiene of the sense of touch. Describe the different *senses*, and give their Anatomy, Physiology and Hygiene. P. 384 to 424. Mention the means of preserving health. P. 425.

CHAPTER XI.
GENERAL QUESTIONS.

SUGGESTION.—The following questions have been used in the *examinations* of *candidates* for *teaching*, in Boston, New York, Cincinnati, Chicago, Madison, Milwaukee, and by several *County School Commissioners*, in different States. Some of these questions you have met before; be careful, therefore, you do not give an answer inconsistent with the one you have given in another place, the first being correct.

1. Define Orthography.

2. Correct the following sentence as to the use of capitals, and give the rules for your corrections:

it is true as i have often heard That a poor speller can never be a successful teacher.

3. Correct the following sentence as to spelling and the use of capitals:

upon the Thirty first day of december a. d. 1861 the tirm of offis of all town Superntendents turminated.

4. Correct the spelling of such words as are misspelled in the following list :

Benefited, Superseded, Monies, Scholar, Truely, Always, Preferred.

5. What is a Prefix? What is a suffix?

6. Give three Prefixes, with their meanings.

7. Give three Suffixes, with their meanings.

8. How many different Prefixes do you find in the following words :

Ignoble, Illegal, Immoral, Inelegant.

9. Separate by a hyphen the Prefixes from the rest of the following words:

Antedate, Induce, Subscribe, Reflect, Suggest, Extraordinary, Describe.

10. Separate by a hyphen the Suffixes from the rest of the following words :

Timely, Consignment, Relaxing, Aggressive, Locality, Generalize.

11. Write the Primitive or Root Word found in the following :

Justify, Ignoble, Unmanly, Using, Referring, Inconstancy, Infancy.

12. How many sounds in the English language?

13. Why is it so difficult to learn to spell the English language correctly ?

14. Why do persons who spell well orally, often fail in writing words correctly ?

15. Separate the following words into syllables :

Animate, Dictionary, Spelling, Alleviate, Timely, Correction.

16. Correct the following two lines in all particulars needing correction :

I was absent from home when the young lady to whom you referred called.

17. How many sounds are represented by the character C ?

18. How many and what sounds are found in the pronunciation of the word example ?

19. What is an elementary sound ?

20. What is a Vocal or Tonic ?

21. What is an Aspirate?

22. What elementary sounds are represented by more than one character?

23. What is a derivative Word?

24. How many and what are the Vowels?

25. How many and what are the Consonants?

26. What is a diphthong?

27. Correct in all particulars needing correction the following:

this association shall be caled the picwickran club and shall have for its members such persons onely as are wiling to make self improvment there first studdy its moto shel be know thy self.

28. Correct the following, if it needs correction, and give your reasons for your corrections:

I can not conceive how any sane man can believe the storys that are so busyly circulated by persons caring not for reputation or caracter.

29. How many sounds has A?

30. How many sounds has O?

31. Write the plural of body. Monkey.

32. Write the singular of dice genera.

33. Write the plural of sheep. Fleece. Scissors.

34. How many elementary sounds are heard in the word thoroughly?

35. What elementary sounds are heard in the word cough?

36. Correct the spelling of such words as are misspelled in the following list:

Procede, Preceed, supercede, succeed, allegance, schollarship, transmitted, addoration, Tennessee, Mississipi, Cincinnatti.

1. Name the different waters bordering on Michigan; upon Pennsylvania; upon Spain; upon Turkey, in Asia.

2. Name five rivers that flow into the Ohio; five that flow into the Mississippi, upon the eastern side; five that flow into the Atlantic ocean, having their origin in the loyal States.

3. Name five mountain chains upon the Eastern Continent, with the position and direction of each.

4. Name five mountain chains of North America, with the position and direction of each.

5. Name the highest mountain peak in the world. The highest in the United States.

6. Trace the shortest water route from Green Bay, Wis., to New Orleans, Louisiana.

7. Trace the shortest route by railroad and steamboat from Madison, Wis., to Boston, Mass., naming at least five important places on the route.

8. What is the shortest route of travel from the capital of Michigan to the capital of Texas.

9. Locate five large cities in New England. Locate five large cities in the Middle States. Locate five large cities in the Seceded States. Locate five large cities in the States lying north of the Ohio river, and east of the Mississippi, without going farther east than Lake Erie.

10. Through what Grand Divisions of the world does the Equator pass?

11. Why are the Tropics situated 23½ degrees from the Equator?

12. How do you account for the change of Seasons?

13. How do you account for the difference in the length of the day at different seasons of the year?

14. Which contains the larger number of square miles, the Eastern or Western Hemisphere?

15. Which has the larger area, Wisconsin or Pennsylvania?

16. How does South America compare with North America in respect to facilities for inland commerce?

17. Name the principal rivers flowing into Lake Michigan.

18. Give the boundaries of Wisconsin.

19. Name the States bordering on the Mississippi in their order, commencing at the northernmost State upon the eastern side and ending with the northernmost State upon the western side.

20. Which of the large lakes of North America form part of the northern boundary of the United States?

21. What are the Meridians?

22. What are parallels of Latitude?

23. What is the reason for the Polar circles being 23½° from the Poles?

24. What evidence have we that the earth is round?

25. What evidence have we that the earth is flattened at the poles?

26. Give the distance in degrees and minutes between the Tropic of Cancer and the Arctic Circle.

27. Give the boundaries of Ohio.

28. On what part of the globe is the line of perpetual snow the highest?

29. Give the latitude and longitude of the South Pole.

30. Name the seven largest gulfs and bays of North America.

31. How is Calcutta situated? Singapore?

32. Name the grand divisions of the land surface of the globe, and give the largest city in each, with its location.

33. Give the boundaries of France and the location of its principal city.

34. How can a vessel of light draught make its way from Charleston, S. C., to Lake Superior?

35. In order to sail from St. Petersburg to Odessa, through what waters would you have to pass?

36. Bound your own County, and give its lakes, rivers, canals and railroads, if any.

37. Where is Aux Cayes, Corocoa, Cienfuegos and Trieste?

38. Give the location of the four largest river valleys in the world.

39. Name the Peninsulas of Europe, and the direction in which they extend.

40. What range of mountains contains the highest peaks?

41. What languages are principally spoken in Brazil, Moldavia, Switzerland, Quebec, and St. Augustine?

42. Classify the States in the Union according to their mineral resources.

43. Mention the natural advantages which a country should have in order to be fertile.

44. What is the origin and direction of the Gulf Stream?

45. What are the advantages of Mountains to a country?

46. What is the face of the country in Vermont, Florida and Egypt?

1. Define Notation and Numeration.

2. Explain the reason of the first figure of a Partial product under the figure of the multiplier.

3. What is the difference between the greatest common divisor and the least common multiple?

4. What is the difference between common, decimal and duo-decimal fractions?

5. Why does multiplying one proper fraction by another give a product less than the multiplicand?

6. How do you reduce fractions to a common denominator?

7. What effect has multiplying by a proper fraction upon the multiplicand?

8. Why do you invert the divisor in division of fractions?

9. How do you reduce a common fraction to a decimal, and why?

10. Perform the work indicated, and give rules: $0.25 \times 175.0 \div 10$.

11. How many sevenths in $\dfrac{1\frac{1}{2}}{2\frac{3}{4}} \times \dfrac{\frac{3}{8}}{\frac{1}{2}} \div 6\frac{1}{2}$?

12. A sends B $1050 for the purchase of goods, allowing B 5 per cent. commission upon the purchase. The amount sent B is to cover both purchase and commission. What will be the value of the goods purchased?

13. Calculate the interest upon $15.75 at 9 per cent. per annum, for 3 years 5 months and 21 days.

14. What principal at interest for 3 years and six months, at 12 per cent. per annum, will amount to $35,500?

15. In what time will $1,000 at interest at 10 per cent. per annum, amount to $1,534.25?

16. At what rate per cent. will the interest of $800 in 1 year 6 mo. 24 days amount to $75.20?

17. Discount a note for $325, due 5 years and 4 months hence, at 9 per cent. per annum. What is the discount? What is the present worth of the note?

18. How much will be paid upon an Insurance Policy for five years, issued by a Mutual Insurance Company, the premium note

being given for 5 per cent., upon $2,750? The assessments made upon premium note are: advance 35 per cent., and several assessments as follows: .03, .05, 4½ and 11.

19. How much more will it cost to insure $3,500 in a Stock Company for 5 years, at a rate of ⅛ per cent. per annum, than in a Mutual Company when the premium note was given for 4½ per cent. for 5 years, and the several assessments upon the premium note amount to 55 per cent.?

20. A merchant sold a bill of goods for $175, gaining 25 per cent. upon the cost; what did the goods cost him?

21. A builds a wall 25 feet long, 4 feet high, and 2½ feet thick, in 10 days of 10 hours each. In how many days could he build a wall 30 feet long, 6 feet high, and 3 feet thick, working 9 hours a day?

22. What is a multiple?

23. Explain the process of dividing ⅞ by ⅝.

24. Analyze the following question: If six were ten, what would 7 and ⅓ be?

25. What is the difference between 25÷.25, and .25÷25?

26. What sum of money may be drawn at a bank on a note of $468, payable in 45 days?

27. When gold is fifty per cent. premium, how much will be received in exchange for $1,000 in paper?

28. Sold a watch which cost me $30, for $35, on a credit of eight months; what did I gain by the bargain, and how much per cent.?

1. Name the parts of Speech.

2. What is a sentence?

3. How many words are necessary in the construction of a simple sentence?

4. What properties have nouns?

5. What properties have verbs?

6. Correct the following, and give reasons for correction:

"I saw him when he done it."

7. Analyze the following sentence:

South Carolina seceded from the Union on the twentieth day of December, in the year 1860.

8. Correct in all particulars needing correction, the following, and give reasons:

Between you and I the trouble lay nearer home.

9. What is the use of interjections? Illustrate by an example.

10. Correct in all particulars needing correction, the following, and give reasons for corrections:

A great variety of reasons are given for the changes, but every one of the members still hold to their opinions.

11. Correct and give reasons:

"His argument was the best of all others."
"Her appearance was better than that of any person I ever saw."

12. How do you distinguish Relative from Interrogative Pronouns?

13. Is the following correct? If not, wherein does its incorrectness consist?

"I hoped to have seen you." *

14. Analyze the following:

"Thou may'st be popular
Perchance but seek not popularity
As motive-spring of any act in thy profession."

15. In two different propositions use the same word as an Adjective and as a Noun.

16. Write five sentences containing errors, point out the errors and tell why they are such.

17. Correct the sentences following that are incorrect

"Who are you looking for?"
"She is the person whom all love."
"Both were unfortunate but neither are to blame."
"Whom do you charg with folly?"

18. Write a compound sentence containing all the Parts of Speech.

19. Analyze the following sentence:

6

"The term of school which has just commenced will close upon the last Friday of March."

20. Correct the following in all respects as to spelling, punctuation, capitals, and construction:

"twas but the day befor chrismas that he went and done a deed which no man has ever seen the like of it."

21. Give the plurals of *Genus, Emphasis*, and *Criterion.*
22. Compare *Happy, Gay, Useful*, and *Golden.*
23. Decline the personal pronouns *Thou* and *She.*
24. Define *Voice* and *Mood.*
25. Correct the following sentences, viz.:

(*a.*) Both this dress and the other is finished, but neither of them set well.

(*b.*) Who was you speaking to previous to my arrival?

(*c.*) He can neither learn easy or speak gramatical.

26. In the last sentence (*c.*) parse the words *neither* and *speak,* and the word *neither* in the sentence (*a.*)

27. How is gender expressed?

28. What class of verbs govern two objective cases?

29. When is the subject of a verb not its nominative?

30. *Such as* I esteem shall be invited.   Parse the words in italics.

31. "John is a noun."   Parse *John,* giving gender, number, person, &c.

32. "A man who is industrious will be respected."   Analyze.

33. "I have purchased an ox, therefore I can not come."   What is the relation of the latter clause to the former?

34. They come to the number of one hundred men.   Parse the clause "to the number of one hundred men."

35. "It is they who deceive you."   Is the sentence correct?   If not, correct it.

36. What words are essential to a sentence?   Form one containing all the parts of speech.

37. Of the two Latin Poets, Virgil and Horace, "the first is the most celebrated."

38. "Ten idle men were collected to see if it were Washington, him whom the loyal citizens honors."   Correct.

39. "I intended to have gone."   Correct and give reasons.

40. "When the cars arrived the policeman arrested the man who stepped upon the platform." Analyze.

41. Correct the following sentences:

"There comes three persons either of which accomplish with ease what you propose."

42. "The farmers men-servaut brought to market turkeys and potatoes which he delivered at Smiths, the tailors servants by the hands of the Messrs. Browns."

43. "I they and you having completed your studies, it becomes us to be as them who all respect for their virtue."

44. What is Prosody?

45. In punctuation, what does the dash denote?

46.   "Awake my St. John, leave all meaner things
         To low ambition and the pride of kings."

47. Tell the *kind* of verse, the *number* of poetic feet, and put the accent on the long syllables.

48. What is the *logical subject*, and what the *grammatical predicate* in a sentence?

49. What is meant by *Declension* in grammar? What by *Inflection*?

50. Give an example of *Personification*. Of Metaphor. Of Simile.

51. In the sentence, "He that glorieth let him glory in the Lord," parse the words He, let, and glory.

52. Correct the sentence,

"He learns me grammar, but neither of us speak English correct."

53. Analyze the sentence,

"A desire to excel will stimulate to exertion."

1. By what Governments of Europe were the earlier settlements made?

2. How many wars were there between the English Colonies and the French and Indians, and how are the three most prominent designated?

3. What were the chief causes of the alienation of the colonies from the English Government?

4. When and where was the first Provincial Congress formed? §

5. When and where was the first blood shed in the war of the Revolution?

6. When did the battle of Bunker Hill take place? *14 June*

7. When did Cornwallis surrender at Yorktown? *19 Oct 1781*

8. When and where was the treaty signed by which Great Britain acknowledged the independence of the United States?

9. When was the Federal Constitution adopted?

10. Who was the fourth President of the United States, and how long did he hold office?

11. In what year did the second war with Great Britain commence, and when did it terminate?

12. Give a history of what is known of the Hartford Convention.

13. With what other nations besides England has the United States been at war?

14. Give the prominent events with the war with Mexico.

15. When was Michigan admitted into the Union?

16. When was Wisconsin admitted into the Union?

17. Who was the first Governor of this State?

18. What noted events occurred in the Territorial history of Wisconsin?

19. Name the four great epochs in U. S. History.

20. When, where, and by whom was the first permanent settlement made in North America?

21. What causes led to the American Revolution?

22. Give the names and dates of the four most important battles of the Revolutionary War, and a brief account of each.

23. Name three of the most noted commanders, and give an account of each.

24. When, where, and by what terms was peace concluded?

25. What led to the war of 1812?

26. Give an account of the *land* operations during this war.

27. Give an account of the *naval* operations during this war.

28. Give the names and dates of the principal Indian wars.

29. How did the U. S. obtain possession of Louisiana?

30. How was the Federal Constitution framed? When did it go into operation?

31. Give the principal events of Washington's Administration.

32. What was the great event of Monroe's Administration?

33. In whose administration occurred the Algerine War?

34. Give a short account of the Mexican War.

35. What Americans have become celebrated for great and useful inventions?

36. Name the most noted naval commanders of the U. States.

37. Give a brief account of Jackson's Administration.

38. Tell what you know of the present Rebellion against the United States.

1. What causes a difference of climate at different points upon the same parallel of latitude?

2. What is the cause of land breezes?

3. What is the cause of sea breezes?

4. Explain the causes of the Trade Winds.

5. Why is the Pacific coast of the U. S. warmer than the Atlantic coast, upon the same parallel of latitude?

6. What are the prominent physical features of North America?

7. What are the oceanic currents?

8. Describe the Gulf Stream.

9. What is the difference between an earthquake and a volcano?

10. What is the cause of Water Spouts?

11. In what respects do the grand divisions of the globe differ from each other?

12. In what respects do the grand divisions of the globe resemble each other?

13. To what causes may we attribute the fertility and productiveness of the Mississippi Valley?

14. Why is Siberia colder than the same latitude of British America?

15. Contrast the vegetable productions of Equatorial Africa and South America.

16. Contrast the animals of tropical and frigid regions.

17. Contrast the coverings of animals of the temperate zones during the summer and winter months.

18. What is the difference between frost and dew?

19. What are the principal causes of rain?

20. What conditions are essential to the production of hail?

21. Name the principal productions of the tropical regions. Of the temperate regions.

22. What is the difference between the soils of New England and of Wisconsin?

23. Why are there no large rivers in Peru?

24. Why does it seldom rain in Egypt?

25. What is the cause of wet and dry seasons in California?

1. Give some account of the first day's work in commencing a school.

2. What is a graded school?

3. How should scholars be classified?

4. How should text-books be used by the scholar?

5. Should the Teacher use a text-book in conducting a recitation?

6. In mental arithmetic should pupils be allowed to use the book in recitation?

7. State the advantages and disadvantages of concert exercises in school.

8. What are essential requisites in the qualifications of good teachers?

9. State some common faults observable in teachers.

10. Should pupils be allowed to report their own delinquencies?

11. To what extent should written records of deportment and scholarship be kept?

12. At what temperature should a school-room be kept?

13. Should giving prizes be encouraged?

14. What advantages and what disadvantages attend giving prizes?

15. What should be the length of recitation required of pupils from four to ten years of age?

16. How many hours per day should children under ten years of age be confined to the school-room?

17. What advantages attend the practice of "boarding round" by teachers?

18. Would you encourage pupils to report the delinquencies of each other?

19. To what extent would you teach morals in school?

20. Would you require "compositions and declamations?"

SECOND PART.

CHAPTER XII.

BOTANY.*

1. Define Botany, and describe the departments into which it is divided. Wood's Class Book, page 13.

2. Describe the relations of Botany to our sustenance, protection, and the healing of our diseases. Wood's Class Book, page 14.

3. Define a plant and give the difference between it and an animal or a mineral; how is it affected by cultivation? Describe the early stages of plant life. 14.

4. Describe each of the elementary tissues that enter into its structure. 20.

5. What are Ducts? Their use? Where found? 23.

6. What is the Epidermis, or Skin, of which it is composed? 24.

7. Describe the Stomata. Give their use and location. 24.

8. What are Hairs, Stings, Glands, Prickles, Thorns? 25.

9. Describe the two grand divisions of the vegetable kingdom, the Phainogamia or Flowering, and the Crytogamia or Flowerless plants. 26.

10. How are they readily distinguished by their Tissues, Seeds, general structure? 27.

11. The Flowering plants are subdivided into Endagenous and Exogenous. Describe the mode of growth and leaf of each. Name example of each class. 77.

12. Name and describe the Floral envelopes. State which constitutes a regular flower. 29, 30.

---

* In the First Grade of all our Grammar Schools, Botany is taught, unless it be postponed to be pursued in the High School. Hence it is expected that Candidates will prepare themselves in this interesting branch.

13. Mention those organs which are essential for the production of fruit. What is the office of the Pollen?

14. What did Linnæus take as the basis of the Artificial System of classification of the Genera? 34.

15. How does it differ from the system of Jussieu? 112.

16. Describe compound and simple Ovaries and the Ovules. 42.

17. What do you understand by Dehiscence? Describe the different modes.

18. What is the ultimate product of vegetation? 57.

19. Describe the parts of the seed. Where is the Embryo plant found? What are its parts? 57, 58.

20. What is the Cotyledom, and what office does it perform to the new plant? 58.

21. What is Germination? What are essential conditions to it? 60.

22. Define a root, and give its office to the plant. 62.

23. In what part of the root does Absorption take place? 67.

24. Describe the different forms of the root. 63, 64.

25. How can you prove that Absorption takes place in the Spongioles?

26. Define the stem and tell wherein it differs from the root. 62.

27. Tell the difference between a Branch, Thorn and Prickle. 71.

28. How does a leaf bud differ from a flower bud? 70.

29. What are Axillory and Terminal buds? 70.

30. Describe the Caulis, Runner, Scape, Vine, Trunk, &c.? 74.

31. How does the Herbaceous stem differ from the Woody? 77.

32. Describe the mode of growth, and the bark, pith, and woody layers of Exogenous stem. 77.

33. How can the age of a tree be ascertained? 78.

34. Describe the mode of increase of the Endogenous stem, and tell what each bundle consists of. 81.

35. What is vernation? Give the different modes of folding the leaf in the bud. 82.

36. When are leaves said to be Opposite, Alternate, Verticellate and Fasciculate? 83.

37. When are leaves said to Cauline? When Radicle? 84.

38. When Net Veined? When Parallel Veined? When are leaves Simple? When Compound? 85.

39. What is the Skeleton and Venation of the leaf? 77.

40. Name and describe the different forms of the Feather Veined leaf, and mention examples of each form. 87.

41. Describe the different forms of Parallel Veined leaves. 89.

42. How are the Margins of leaves modified by the Venation? Describe the forms of Margins. 90.

43. When is the Apex of the leaf said to be Entire? Dentate, Serrate, Crenate, Spinous, Lacinate, &c.? 90.

44. When is the surface of the leaf Rough, Pubescent, Glabrous, Pilose, Vilose, Rugose, Woody, Hoary?

45. Describe the Compound leaf and name its parts. 91. When is the leaf Pinnate, Bipinnate, Tripinnate?

46. When is the leaf Amplexicaul, Perfoliate, Connate, &c.? 93.

47. Describe the Sarracenia or Pitcher plant. 93.

48. When are leaves Deciduous, Fugacious, Persistent? 96.

49. Describe Exhalation, Absorption, Respiration. 98.

50. How can you illustrate by experiment? 99.

51. Define Digestion in plants. Tell where and how performed. 101.

52. What is Inflorescence? Describe the different modes. 102.

53. Mention some of the Chemical elements that enter into the structure of plants. 106.

54. Mention any other important principles in Botany.

CHAPTER XIII.

ALGEBRA.

REMARK.—In examinations in Algebra it has been found that the majority of students have devoted their energies mainly to the solution of problems, carelessly passing over the *principles* involved in the questions proposed. This is a fatal error. Every Candidate should carefully prepare himself in the *definitions*, and thoroughly qualify himself in the *principles* of the *Science*; then all problems can be easily solved.

In giving answers give *reasons* for every statement made, whether called for or not.

NOTE.—The references in the following questions in Algebra are to "Davies' Bourdon." P. stands for page. Art. for article. Ex. for example.

1. What do you understand by Quantity? Art. 1.
2. What is Mathematics? Art. 2.
3. Define Algebra. Art. 3.
4. How many kinds of quantities are considered in Algebra? Art. 4.
5. Name and describe those quantities. Art. 4.
6. How many *signs* are used in Algebra?
7. What is the sign for Addition, and how made? Art. 5.
8. Make the sign for Subtraction, and tell its meaning. Art. 6.
9. Which is the Positive Sign? Which Negative? Art. 6.
10. How many signs are there for Multiplication? Art. 7.
11. Make the signs for Multiplication. Art. 7.
12. How many are there for Division? Art. 8.
13. Make signs for Division. Art. 8.
14. Make and define the sign of Equality. Art. 9.
15. Define the sign of Inequality. Make it. Art. 10.
16. What sign is used to denote that there is a difference between two quantities without knowing which is the greater? Art. 11.
17. What sign is used to denote that one quantity varies as another? Art. 12.
18. What are the signs of Proportion? How read? Make them. Art. 13.
19. What sign is used to denote *hence* or consequently? Art·

20. What is a Coefficient?   Give an example.

21. When no coefficient is expressed, what is understood? Art. 14.

22. What is an Exponent, and what does it show?   Art. 15.

23. When no exponent is written, what is understood?   Art. 15.

24. What is the *Power* of a quantity?   Degree of a quantity? Art. 16.

25. What relation between the exponent and the number of Multiplications?   Art. 16.

26. Illustrate the use of the exponent by taking a as a factor six times, b eight times, c seven times.   P. 18.

27. What is the Root of a quantity?   Art. 18.

28. What is the *Radical Sign?*   Make it.   Art. 18.

29. Give an example using the *Radical Sign.*   Art. 18.

30. What is the reciprocal of a quantity?   Art. 19.

31. Define an *Algebraic quantity.*   Art. 20.

32. Give an example to illustrate Algebraic quantities.   Art. 20.

33. What is a *monomial* or *term?*   Art. 21.

34. Define a Binomial.   Trinomial.   Polynomial.   Art. 21.

35. What is the *Numerical value* of an Algebraical expression? Art. 22.

36. What is an additive term?   Subtractive?   Art. 23.

37. What effect does changing the order of the terms of a polynomial have on the *numerical value* of the quantity?   Art. 24.

38. Define the Dimension of a term.   Art. 25.

39. What are the literal factors of a term?   Art. 25.

40. How do you tell the degree of a term?   Art. 25.

41. When is a polynomial Homogeneous?   Art. 26.

42. Write a polynomial that is homogeneous.   Art. 26.

43. What is a vinculum?   Parenthesis?   Brackets?   Art. 27.

44. Make the characters named in the last question.   Art. 27.

45. What are Similar terms?   Art. 28.

46. Write terms that are similar, and those that are dissimilar. Art. 28.

47. When is a polynomial reduced to its simplest form? Art. 29.

48. Give the *Rule* for reducing a polynomial to its simplest form. Art. 29.

49. In reducing a polynomial what effect does it have on the coefficients and exponents? P. 23.

50. Define a Theorem and a Problem. Arts. 30, 31.

51. Define a Formula? P. 25.

52. Solve the following, and give the formula for it. Art. 31.

53. The sum of two numbers is 67, and their difference is 19; what are the numbers? Art. 31.

54. The sum of two numbers is a, and their difference b; what are the numbers? Art. 31.

55. Give a *formula* involving the principle of the last example. P. 25.

ADDITION.

1. Define Addition. Art. 31.

2. When the quantities are dissimilar how do you add them? Art. 32.

3. Give the *Rule* for the addition of Algebraic quantities. Art. 34.

SUBTRACTION.

1. Define Subtraction. Art. 35.

2. When the quantities are *not similar*, how do you subtract? Art. 36.

3. Give the *Rule* for subtraction of algebraic quantities. Art. 37.

4. Give the *reasons* for changing the signs of the subtrahend. Art. 37.

5. If you have an algebraic quantity within a parenthesis, and a *minus* sign before it, what effect does it have on the terms when the parenthesis is omitted? Art. 38.

6. Illustrate the last question by an example. P. 32.

7. Do the words *add* and *sum* always mean augmentation? Art. 39.

8. Explain the difference between an Arithmetical and Algebraic Sum. Art. 39.

9. Do the words subtraction and difference always mean diminution? Art. 39.

10. Are the algebraic signs, *plus* and *minus*, always the true signs of the *terms* before which they are placed? Art. 40.

11. Illustrate the last question by an example. Art. 40.

MULTIPLICATION.

1. What is multiplication in Algebra? Art. 41.

2. Name and explain the terms used in multiplication. Art. 41.

3. What is the RULE for multiplication of monomials? P. 34.

4. How do you multiply one polynomial by another? Arts. 43–45.

5. Multiply a—b by c—d, and give the reasons for every step. Art. 44.

6. How can you make it appear that minus multiplied by minus gives plus? Art. 43.

7. In multiplication, when *both factors* are *homogeneous*, how will the product be? Art. 46.

8. How many terms will there be in the product if no *two terms* of the product are similar? Art. 46.

9. Among the terms of the product, how many terms will there always be which can not be reduced with any others? Art. 46.

10. Name the terms intimated in the last question, and tell why. Art. 46.

11. Give the *formula* for the square of the sum of two quantities. Art. 47.

12. State the *formula* for the square of the difference of two quantities. Art. 47.

13. Give the formula for the *sum* of two quantities multiplied by their *difference*. Art. 47.

14. What is the *law* of the product of two quantities? Art. 48.

DIVISION.

1. What is Division? Art. 49.

2. Name the terms used in Division, and define them. Art. 49.

3. How do you divide one monomial by another monomial? Art. 51.

4. State the principle in regard to the signs in division. Art. 50.

5. State two cases in which the *exact division* of monomials is impossible. Art. 52.

6. State the principle in regard to the exponents of the dividend and divisor. Art. 53.

7. Show that any quantity whose exponent is 0, is equal to 1. Art. 54.

8. How do you divide a polynomial by a monomial? Art. 55.

9. How do you divide one polynomial by another? Art. 56.

10. What do you understand by arranging the dividend and divisor with reference to a certain letter? Art. 56.

11. Give the reasons for the whole process in the division of · polynomials. Art. 56.

12. When is the exact division of one polynomial by another impossible? How many cases are there? Art. 58.

FACTORING POLYNOMIALS.

1. What do you understand by Factoring Polynomials? Art. 59. ·

2. How may a polynomial be resolved into two or more factors? Art. 59.

3. Find the factors of the following: n^3+2n^2+n. Ex. 5. P. 52.

4. Find the factors of the following: a^2x-x^3. Ex. 7. P. 52.

5. Demonstrate the following proposition: The difference of the same powers of any two quantities is exactly divisible by the difference of the quantities. Art. 60.

6. Illustrate the last by the following: Divide a^m-b^m by $a-b$. Art. 60.

7. Demonstrate the following: The sum of the odd powers of the same degree of two quantities is always divisible by the sum of the quantities.

8. Divide (a^3+b^3) by $(a+b)$, and give the reasons.

9. Give the *rule* and *reason* for finding the Greatest common divisor of two or more polynomials.

10. Find the *greatest common Divisor* of the following example: $2x^4+11x^3-13x^2-99x-45$ and $2x^3-7x^2-46x-21$.

11. Give the Rule and reason for finding the *Least Common Multiple* of two or more polynomials.

· 12. Find the least common multiple of the following example: $3x^2-11x+6$, $2x^2-7x+3$, and $6x^2-7x+2$.

ALGEBRAIC FRACTIONS.

1. What is an Algebraic Fraction? Art. 62.

2. Define a *Fractional Unit.* Art. 62.

3. What are the *Terms* of a fraction? Art. 63.

4. What is an Entire quantity? Art. 63.

5. What is a mixed quantity? Art. 63.

6. When may the Fraction be reduced to an entire quantity? Art. 64.

7. What effect does multiplying the Numerator of a fraction have upon the value of the fraction? Art. 65.

8. What effect does multiplying the denominator of a fraction have upon its value? Art. 66.

9. What effect does multiplying both numerator and denominator have upon the value of the fraction? Art. 67.

10. What effect has dividing both numerator and denominator of a fraction have upon its value? Art. 67.

11. Give reasons for your answers to the last three questions. P. 56.

12. How do you reduce a fraction to its lowest form?

13. Reduce $\dfrac{18a^2c^2-3acp}{27ac^2-6ac^3}$ to its simplest form. Ex. 7. P. 57.

14. How do you reduce a *mixed* quantity to a fractional form? P. 57.

15. Reduce $3x-1-\dfrac{x+a}{3a-2}$ to the form of a fraction. Ex. 6. P. 58.

16. Why do you change the signs of the terms of the numerator in the last example? Art. 38.

17. How do you reduce a *fraction* to an *entire* quantity? R. P. 58.

18. Reduce $\dfrac{10x^2-5x+3}{5x}$ to a mixed quantity. Ex. 6. P. 59.

19. Reduce $\dfrac{x^3-y^3}{x-y}$ to an entire quantity. Ex. 5. P. 59.

20. How do you reduce fractions having different denominators to equivalent fractions having a common denominator? R. P. 60.

21. Reduce $\dfrac{a}{a-b}$, $\dfrac{c-b}{ax}$ and $\dfrac{b}{c}$, to equivalent fractions having common denominators. Ex. 6. P. 60.

22. How do you Add fractions? R. P. 61.

23. What is the sum of $\dfrac{a-x}{a-b}$, $\dfrac{c}{a+b}$ and $\dfrac{d}{a+x}$? Ex. 10. P. 62.

24. Give the process for the *subtraction* of fractions. R. P. 62.

25. From $3x+\dfrac{x}{b}$ take $x-\dfrac{x-a}{c}$. Ex. 7. P. 63.

26. How do you *multiply* one fractional quantity by another? R. P. 62.

27. Multiply $a+\dfrac{ax}{a-x}$ by $\dfrac{a^2-x^2}{x+x^2}$. Ex. 8. P. 64.

28. How do you *divide* one fraction by another? R. P. 65.

29. Divide $\dfrac{x^4-b^4}{x^2-2bx+b^2}$ by $\dfrac{x^2+bx}{x-b}$. Ex. 8 P. 66.

30. Divide $\dfrac{a+1}{a-1}$ by $\dfrac{1+a}{1-a^2}$. Ex. 10. P. 66.

31. What effect will it have on the quotient to change the signs either of the numerator or denominator? Art. 69.

32. How will the value of the fraction be affected by adding the same quantity to *both terms* of a *proper* fraction? Art. 70.

33. By adding the same quantity to both terms of an improper fraction? Art. 70.

34. Demonstrate the principle in the last two questions.

35. If the same quantity be subtracted from each term of a proper fraction, how will the value of the fraction be affected? Art. 70.

36. By subtracting the same quantity from each term of an

improper fraction, what effect on the value of the fraction? Art. 70.

37. Explain the principle in the last two questions.

38. Multiply $\dfrac{x^2-9x+20}{x^2-6x}$ by $\dfrac{x^2-13x+42}{x^2-5x}$ and get $\dfrac{x^2-11x+28}{x^2}$. Ex. 5. P. 68.

39. Divide $1+\dfrac{n-1}{n+1}$ by $1-\dfrac{n-1}{n+1}$ and get n. Ex. 8. P. 68.

40. From $\dfrac{1+x^2}{1-x^2}$ take $\dfrac{1-x^2}{1+x^2}$. Ex. 4. P. 68.

41. What does the sign *Zero* signify? Art. 71.

42. What is the sign of *infinity*? Art. 71.

EQUATIONS OF THE FIRST DEGREE.

1. What is an Equation? Art. 72.

2. What are members of an Equation? Art. 72.

3. What is the First Member? Which the Second? Art. 72.

4. How many unknown quantities may an equation have? Art. 73.

5. How are equations classified? Art. 73.

6. How can you tell what *degree* an Equation is? Art. 73.

7. What are Numerical Equations? Art. 74.

8. Define *Literal* equations. Art. 74.

9. What is an *Identical equation?* Art. 75.

10. State the *properties* of an equation. Art. 76.

AXIOM.

1. Define an Axiom. Art. 76.

2. How many axioms are used in Algebra? Art. 76.

3. Give the six axioms. Art. 76.

SOLUTION OF EQUATIONS.

1. What do you understand by the Solution of an equation? Art. 77.

2. What do you understand by the Transformation of an Equation? Art. 78.

3. Of what does the First Transformation consist? Art. 78.

4. How do you transform an equation involving fractional terms to one involving only entire terms? R. P. 76.

5. Reduce $\dfrac{ax}{b} - \dfrac{2c^2x}{ab} + 42 = \dfrac{4bc^2x}{a^3} - \dfrac{5a3}{b^2} + \dfrac{2c^2}{a} - 3b$, to an equation involving only entire terms. Ex. 4. P. 77.

6. Of what does the second transposition consist? Art. 79.

7. How do you transpose a *term* of an equation from one *member* to the other? R. P. 77.

8. Upon what principle is the Rule founded for the last question?

9. Give the *Rule* for *Solving* an *equation* of the first degree. R. P. 78.

10. Find the value of x in the following: $2x - \dfrac{4x-2}{5} = \dfrac{3x-1}{2}$. Ex. 16. P. 80.

11. Solve the following: $\dfrac{(a+b)(x-b)}{a-b} - 3a = \dfrac{4ab-b^2}{a+b} - 2x + \dfrac{a^2-bx}{b}$. Ex. 18. P. 80.

12. Of how many parts does the solution of a problem consist? Name them. Art. 81.

13. Of what does the statement consist? Solution? Art. 81.

14. What is the *Rule* for "Stating" problems? R. P. 81.

15. Solve the following: A capitalist receives a yearly income of $2940; four-fifths of his money bears an interest of 4 per cent., and the remainder of five per cent.; how much has he at interest? Ex. 18. P. 87.

16. In a certain orchard one-half are apple trees, one-fourth peach trees, one-sixth plumb trees, 120 cherry trees, and 80 pear trees; how many trees in the orchard? Ex. 20. P. 87.

17. A person in play lost one-fourth of his money, and then won 3 shillings; after which he lost one-third of what he then had; and this done, found that he had but 12 shillings remaining: what had he at first? Ex. 28. P. 88.

ELIMINATION.

1. Define Elimination. Art. 83.

2. How many methods of Elimination are there? Art. 80.

3. Give the method by Addition and Subtraction. R. P. 91.

4. Explain the method by Substitution. Art. 85.

5. Illustrate by an example the method of Elimination by comparison. Art. 86.

6. How do you solve a problem involving three equations and three unknown quantities? Art. 87.

7. What is a Simultaneous equation? Art. 82.

8. Give the general *Rule* for solving a problem containing any number of equations and unknown quantities. R. P. 94.

9. Given $2x+3y=16$ and $3x-2y=11$, to find the values of x and y. Ex. 1. P. 95.

10. $\frac{x}{7}+7y=99$, and $\frac{y}{7}+7x=51$, to find the values of x and y. Ex. 3. P. 95.

11. Given $7x-2z+3u=17$. $4y-2z+t=11$. $5y-3x-2u=8$. $4y-3u+2t=9$, and $3z+8u=33$, to find the values of x, y, z, u, and t. Ex. 8. P. 95.

12. Solve the following: A's age is double B's, and B's is triple C's, and the sum of their ages is 140; what is the age of each? Ex. 11. P. 99.

13. A footman agreed to serve his master for £8 a year and a livery, but was turned away at the end of 7 months, and received only £2 13s 4d and his livery; what was its value? Ex. 16. P. 100.

14. If A and B together can perform a piece of work in 8 days, A and C together in 9 days, and B and C in 10 days, how many days would it take each person to perform the same work alone? Ex. 20. P. 100.

15. A banker has two kinds of money; it takes a pieces of the first to make a crown, and b pieces of the second to make the same sum. Some one offers him a crown for c pieces. How many of each kind must the banker give him? Ex. 28. P. 102.

INDETERMINATE EQUATIONS AND PROBLEMS.

1. Define an *Indeterminate Equation.* Art. 88.

2. What is an Indeterminate Problem? Art. 88.*

3. How many equations *must* there be for a given number of unknown quantities?

4. What do you understand by the Interpretation of Negative Results? · Art. 89.

5. Solve and explain the following: A Father has lived a number of years expressed by a; his son a number of years expressed by b. Find in how many years the age of the son will be one-fourth the age of the father. Ex. 2. P. 107.

6. State the four principles in regard to negative results. P. 108, 109.

7. What do you understand by the *Discussion of Problems?* Art. 91.

8. What is an Arbitrary quantity? Art. 91.

9. Give and solve the problem of the Couriers. Art. 91.

10. Explain all the *conditions* of the last question. Art. 91.

INEQUALITIES.

1. What is an Inequality? Art. 92.

2. State the *six distinct principles* belonging to inequalities. P. 114, 115, 116.

3. Find x in the following: $\frac{bx}{7} - ax + ab < \frac{b}{7}$. · Ex. 5. P. 116.

POWERS AND ROOTS.

1. What is the square of a quantity? Art. 93.

2. Define the Square Root of a quantity. Art. 93.

3. The square of a Number composed of tens and units is equal to what? Art. 94.

4. Illustrate the last question by squaring 64.

5. Also by squaring 365.

6. Extract the square Root of 96785436.

7. How do you extract the square root of a number? Art. 95.

8. Demonstrate the *Rule* for *square root.* Art. 95.

9. When can you increase the entire part of the root by 1? Art. 95. P. 122.

10. To what is the number of places in the root always equal? Rem. II. P. 123.

11. Is the square root of an imperfect square commensurable with 1. Rem. 3. P. 123.

EXTRACTION OF THE SQUARE ROOT OF FRACTIONS.

1. How do you extract the square root of a fraction? Art. 96.

2. How do you extract the square root of a fraction when the numerator and denominator are not both perfect squares? Art. 96.

3. How do you extract the square root of a whole number which is an imperfect square to within less than a given fractional unit? Art. 97.

4. How do you obtain the approximate root in decimals? Art. 97.*

5. Give the *rule* for extracting the square root of a vulgar fraction in terms of a decimal. Art. 99.

6. Find the $\sqrt{2\frac{1}{3}}$ to within less than 0.0001. Ex. 2. P. 129.

EXTRACTION OF THE SQUARE ROOT OF ALGEBRAIC QUANTITIES.

1. How do you extract the square root of Monomials? Art. 100.

2. How do you extract the square root of a Polynomial? Art. 101.

3. Demonstrate the Rule for square root of polynomials. Art. 101.

4. Find the square root of $4x^6+12x^5+5x^4-2x^3+7x^2-2x+1$. Ex. 4. P. 132.

5. Find the square root of $25a^4b^2-40a^3b^3c+76a^2b^2c^2-48ab^2c^3$ $+36b^2c^4-30a^4bc+24a^3bc^2-36a^2bc^3+9a^4c^2$. Ex. 6. P. 132.

6. Is a binomial a perfect square? P. 133.

7. When is a trinomial a perfect square? P. 133.

8. $\sqrt{9a^6-48a^4b^2+64a^2b^4}=$ what? P. 133.

RADICAL QUANTITIES OF THE SECOND DEGREE.

1. Define a Radical Quantity. Art. 102.

2. What is a Radical of the *third degree?*

3. Define Similar Radicals. Art. 103.

4. How do you simplify a Radical of the second degree? Art. 105.

5. Give the two principles upon which the simplification of radicals depend. Art. 104.

6. How do you Add Radicals? Subtract Radicals? Art. 106.

7. How do you Multiply Radicals? Art. 107.

8. How do you Divide one Radical by another? Art. 108.

9. Give the sum of $\sqrt{\frac{2}{3}}$ and $\sqrt{\frac{27}{80}}$. Ex. 7. P. 142.

10. Give the sum of $\frac{2}{3}\sqrt{\frac{1}{8}}$ and $\frac{2}{3}\sqrt{\frac{7}{10}}$. Ex. 12. P. 142.

EQUATIONS OF THE SECOND DEGREE.

1. Define an equation of the second degree. Art. 110.

2. Write out the form to which every equation of the second degree may be reduced. Art. 111.

3. What does an incomplete equation of the second degree involve? Art. 112.

4. How many roots has every incomplete equation of the second degree? Art. 113. P. 145.

5. Give the *Rule* and reason for solving an equation of the second degree. Art. 114. P. 147.

6. Find x in the following: $mx^2+mn=2m\sqrt{nx}+nx^2$. Ex. 8. P. 151.

7. Find x in the following: $a^2+b^2-2bx+x^2=\dfrac{m^2x^2}{n^2}$. Ex. 15. P. 151.

8. What number is that which being divided by the product of its digits, the quotient will be 3? and if 18 be added to it the order of its digits will be reversed? Ex. 6. P. 154.

9. What two numbers are those whose difference is 15, and of which the cube of the lesser is equal to half their product? Ex. 10. P. 155.

10. Two partners, A and B, gained $140 in trade. A's money was 3 months in trade, and his gain was $60 less than his stock; B's money was $50 more than A's, and was in trade 5 months: what was A's stock? Ex. 11. P. 155.

11. Give the four forms in which an Equation of the second degree may be expressed. Art. 117.

12. Give and solve the Problem of the Lights, with all its conditions. Art. 121.

13. Given $\dfrac{\sqrt{x+a}}{x}+2\sqrt{\dfrac{a}{x+a}}=b^2\sqrt{\dfrac{x}{x+a}}$, to find x. Ex. 4. P. 166.

14. Given $\dfrac{\sqrt{x}+\sqrt{x-a}}{\sqrt{x}-\sqrt{x-a}}=\dfrac{n^2a}{x-a}$, to find x. Ex. 6. P. 167.

15. Given $\dfrac{a+x+\sqrt{2ax+x^2}}{a+x}=b$, to find x. Ex. 8. P. 167.

TRINOMIAL EQUATIONS.

1. Define a Trinomial Equation. Art. 122.

2. Give the form to which every Trinomial Equation may be reduced. Art. 129.

3. Give the *Rule* for solving a trinomial Equation. R. P. 168.

4. What does the solution of a Trinomial Equation of the Fourth degree require? Art. 125.

5. Reduce the following to its simplest form :-

$$\sqrt{ab+4c^2-d^2-2\sqrt{4abc^2-abd^3}}.\quad \text{Ex. 7. P. 172.}$$

6. Given $x^2+x+y=18-y2$ and $xy=6$, to find x and y. Ex. 9. P. 178.

7. The sum of two numbers is 8, and the sum of their cubes is 152; what are the numbers? Ex. 6. P. 182.

8. What two numbers are those whose sum multiplied by the greater is equal to 77, and whose difference multiplied by the lesser is equal to 12? Ex. 9. P. 182.

9. Divide 100 into two such parts that the sum of their square roots may be 14. Ex. 10. P. 182.

10. Two merchants sold the same kind of stuff; the second sold 3 yards more of it than the first, and together they received 35 dollars. The first said to the second, "I would have received 24 dollars for your stuff." The other replied, "And I would have received $12\frac{1}{2}$ dollars for yours." How many yards did each of them sell? Ex. 18. P. 183.

11. Given $(x^6+1)y=(y^2+1)x^3$, and $(y^6+1)x=9(x^2+1)y^3$, to find x and y.

PERMUTATIONS, ARRANGEMENTS AND COMBINATIONS.

1. Define Permutations, and give the *Law* governing them. Art. 130.

2. Define Arrangements, and give the *law* governing them. Art. 131.

3. Define Combinations, and give the *law* governing them. Art. 132.

BINOMIAL THEOREM.

1. What is the Binomial Theorem? Art. 134.

2. Explain and give reasons for the Binomial Formula. Art. 135.

3. What is the *law* for the coefficients and exponents? Art. 137. 138.

EXTRACTION OF ROOTS.

1. How do you extract the cube root of a number. Art. 141.

2. How do you extract any root of numbers? Art. 142.

3. How do you extract any root of Algebraic quantities? Art. 147, 148.

4. Explain the *principles* governing the Transformation of a radical of any degree. Art. 150–159.

5. Explain the *Rules* for *imaginary expressions.* Art. 162, 163.

6. Explain the *principles* governing Fractional and Negative Exponents. Art. 164–170.

ARITHMETICAL PROGRESSION.

1. Of what does a series consist? What is Arithmetical Progression? Art. 171, 172.

2. How do you find the *sum* of the *terms* of an Arithmetical Progression? Art. 176.

3. Explain the *Formulas* belonging to Arithmetical Progression. Art. 176.

4. Find 9 Arithmetical means between each antecedent and consequent of the progression 2.5.8.11.14. Ex. 7. P. 241.

GEOMETRICAL PROGRESSION.

1. Define Geometrical Progression, and give the *rules* for it.

2. Explain and give reasons for the Geometrical Formulas. Art. 187–192.

3. Explain the principle of Indeterminate Co-efficients. Art. 193–198.

4. Explain the Principle of Recurring Series. Art. 199–201.

5. Give the General Demonstration of the Binomial Theorem. Art. 202.

6. State the principles governing the Summation of Series. Art. 208.

7. Explain the principles of Piling Balls. Art. 210–214.

8. How many balls in an incomplete oblong pile, the numbers in the lower courses being 92 and 40, and the numbers in the corresponding top courses being 70 and 18? Ex. 7. P. 274.

9. Explain continued Fractions and Exponential quantities. Art. 215–224.

10. What are Logarithms? Give the General Properties of them. Art. 227–229.

11. Demonstrate clearly the Principles of Logarithms. Art. 230–241.

12. How do you calculate simple and compound Interest by Algebraic Formulas? Art. 245.

13. Give and explain the General Theory of Equations. Art. 244–250.

14. Demonstrate the General Principle of Elimination. Art. 270.

15. Demonstrate the principles for finding the Greatest Common Divisor. Art. 252–261.

16. State the principles involved in the solution of numerical equations containing but one unknown quantity. Art. 275–280.

17. Explain the principles governing the Limits of Positive Roots. Art. 281–285.

18. Explain Descartes' Rule. Art. 293.

19. Give and explain Sturm's Theorem. Art. 298–307.

20. Find the roots of the equation $x^5 - 2x^3 + 1 = 0$. Ex. 5. P. 378.

21. Explain Cardan's Rule for solving cubic equations. Art. 308.

22. What are the roots of the equation $x^3 - 7x^2 + 14x = 20$. Ex. 3. P. 381.

23. Give the Preliminaries to *Horner's method*. Art. 309.

24. Explain the Principles involved in Horner's method. Art. 300–314.

25. Find the roots of the equation $x^5 - 10x + 6 + 1 = 0$. Ex. 4. P. 400.

NOTE.—Many additional questions might have been proposed, but candidates who can answer satisfactorily the above questions need feel no embarrassment in an examination in this science.

CHAPTER XIV.

Every Teacher of our First Class Schools is now expected to be well qualified in the Higher Mathematics. Hence a few questions are proposed in

GEOMETRY.

The references in the questions on Geometry are to "Davies' Legendre." B. stands for Book. D. for Definition. P. for Proposition. C. for Corollary.

1. Define Geometry.

ANS.—Generically it means the art of measuring the earth; but as it is now used Geometry denotes the Science of magnitude in general,—the mensuration of lines, surfaces, solids, with their various relations.

2. Define Extension. Def. 1. B. I.

3. What is a *Point? Line? Straight line?* B. I. D. 5, 6, 7.

4. What is a Broken line? Curved line? Surface? Plane? B. I. D. 8–11.

5. What is a Curved Surface? Plane Angle? Right-Angle? B. I. D. 12–14.

6. What are Oblique Angles? How many kinds? B. I. D. 15.

7. Define an Acute Angle. An Obtuse Angle. B. I. D. 15.

8. When are lines parallel? What is a plane figure? B. I. D. 16, 17.

9. What is a Polygon? Triangle? Hexagon? Octagon? B. I. D. 19.

10. Define an Equilateral polygon. Equiangular polygon. B. I. D. 20.

11. When are two polygons mutually equilateral and equiangular? B. I. D. 22.

12. How are Triangles classified? How many classes are there? B. I. D. 23.

13. Define a Scalene Triangle. An Isosceles triangle. B. I. D. 23.

14. Define Equilateral and Acute angled triangles. B. I. D. 23.

15. Define Right-angled triangles, and obtuse-angled triangles. B. I. D. 23.

16. What are Quadrilaterals? Divided into how many classes? B. I. D. 24.

17. Define Trapezium. Trapezoid. Parallelogram. B. I. D. 24.

18. Into how many classes are parallelograms divided? B. I. D. 25.

19. Define a Rhombus. Rectangle. Square. B. I. D. 25.

20. What is a Diagonal? A Base? B. I. D. 26, 27.

DEFINITION OF TERMS.

1. What is an axiom? Demonstration? Theorem? B. I. D. 27.

2. What is a Problem? Lemma? Proposition? B. I. D. 27.

3. What is a Corollary? Scholium? Hypothesis? Postulate? B. I. D. 27.

EXPLANATION OF SIGNS.

REMARK.—The explanation of the signs in Geometry is the same as is found in the questions on Algebra, which see.

AXIOMS.

1. How many axioms are there? B. I. P. 19.

2. Give them all accurately. B. I. P. 19.

3. How many Postulates are there? Give them. B. I. P. 20.

THEOREMS.

1. Demonstrate Proposition I. Theorem. B. I.

2. Demonstrate Prop. IX. B. I. and P. XXV. B. I. Also Prop. XXVIII. B. I.

OF RATIOS AND PROPORTIONS. B. II.

1. Define Proportion. Ratio. Antecedent. Consequent. B. II. D. 2.

2. How may the ratio of Magnitudes be expressed? B. II. D. 3.

3. When are magnitudes commensurable? When incommensurable? B. II. D. 3.

4. How will you illustrate the principles found in the last two questions? B. II. D. 4, 5.

5. When are four quantities in Proportion? B. II. D. 6.

6. When is a quantity a fourth proportional to the other three? B. II. D. 7.

7. When are *three* quantities in proportion? B. II. D. 8.

8. When are magnitudes in proportion by Alternation? B. II. D. 9.

9. When are magnitudes in proportion by Inversion? B. II. D. 10.

10. When by Composition? When by Division? B. II. D. 11, 12.

11. What are Equimultiples of two quantities? B. II. D. 13.

12. When are two varying quantities reciprocally proportional? B. II. D. 14.

13. Demonstrate P. I, IX and XII. B. II.

OF THE CIRCLE. B. III.

1. Define a *Circle.* Circumference. D. 1. B. III.

2. What is the Radius? Diameter? How do all the radii of equal or the same circles, compare in magnitude? D. 2. B. III.

3. What is an Arc? Chord? Sector? Segment? D. 3, 4, 5. B. III.

4. When is a straight line said to be inscribed in a circle? D. 6. B. III.

5. Define an inscribed triangle. D. 7. B. III.

6. What is an inscribed polygon? D. 7. B. III.

7. Define a Secant line. A Tangent. D. 8, 9. B. III.

8. What is the point of contact? D. 9. B. III.

9. Define the point of tangency. When is a circle inscribed in a polygon? D. 11. B. III.

10. Demonstrate P. IV, VIII, XV and XVIII, in B. III.

11. Demonstrate Problems III, X, XIII and XV, in B. III.

BOOK IV.

1. Define Similar Polygons. D. 1. B. IV.

2. What are homologons, angles and sides? D. 2. B. IV.

3. What do you understand by *area?* Equivalent figures? D. 4. B. IV.

4. When are two sides of one polygon said to be reciprocally proportional to two sides of another? D. 5. B. IV.

5. What are similar Arcs, sectors, or segments? D. 6. B. IV.

6. What is the Altitude of a triangle? D. 7. B. IV.

7. What is the Altitude of a parallelogram? Of a Trapezoid? D. 8, 9. B. IV.

8. Demonstrate the following Proposition:

"The square described on the sum of two lines is equivalent to the sum of the squares described on the lines, together with twice the rectangle contained by the lines." P. VIII. B. IV.

"The square described on the hypothenuse of a right-angled triangle is equivalent to the sum of the squares described on the other two sides." P. XI. B. IV.

"In every quadrilateral inscribed in a circle, the rectangle of the two diagonals, is equivalent to the sum of the rectangles of the opposite sides taken two and two." P. XXXIII. B. IV.

9. Demonstrate Problems 10, 16 and 18. B. IV.

BOOK V.

1. What is a regular polygon? D. 1. B. V.

2. How many sides may a regular polygon have? D. 2. B. V.

3. Demonstrate the following Proposition:

"To inscribe a square in a given circle." P. III. B. V.

4. And the following:

"In a given circle to inscribe a regular decagon." P. VI. B. V.

5. Also this Theorem:

"The arc of a circle is equal to the product of the radius by the circumference." P. XV. B. V.

BOOK VI.

1. When is a straight line perpendicular to a plane? D. 1, 3. B. VI.

2. When is a plane perpendicular to a line? D. 2. B. VI.

3. When are two planes parallel to each other? D. 3. B. VI.

4. Define a diedral angle, and the faces and edge of an angle. D. 4. B. VI.

5. What is the measure of a diedral angle? D. 4. B. VI.

6. Define a Polyedral angle. What is the face, edge and vertex of the Polyedral angle? D. 5. B. VI.

7. Demonstrate the following:

"Two planes which are perpendicular to the same straight line are parallel to each other." P. IX. B. VI.

8. "If two straight lines be cut by three parallel planes, they will be divided proportionally." P. XV. B. VI.

9. "The sum of either two of the plane angles which include a triedral angle is greater than the third." P. XIX. B. VI.

BOOK VII.

1. Define a Polyedron Prism. Base of the prism. D. 1, 2. B. VII.

2. Describe the convex surface of a prism. D. 3. B. VII.

3. Define the altitude of a prism. What is a right prism? D. 5. B. VII.

4. What is a triangular prism? Parallelopipedon? D. 7. B. VII.

5. What is a pyramid? Altitude of a pyramid? D. 9. B. VII.

6. Define a Right Pyramid. Slant Height, Truncated Pyramid. D. 13. B. VII.

7. What is the altitude of a frustrum? Slant Height? D. 14. B. VII.

8. Define the diagonal of a polyedron. Similar polyedrons. D. 16. B. VII.

9. What is a regular polyedron? Homologous parts of a polyedron? D. 18. B. VII.

10. Demonstrate Prop. IV. B. VII:

"The convex surface of a right pyramid is equal to the perimeter of its base multiplied by half its slant height."

11. Two triangular pyramids, having equivalent bases and equal altitudes, are equivalent, or equal in value. P. XV. B. VII.

12. Two similar pyramids are to each other as the cubes of their homologous edges. P. XX. B. VII. Give the general Scholiums to this Theorem.

BOOK VIII.

1. Define a Cylinder, Cone, Sphere, and Spherical Sector. D. 1. B. VIII.

2. Define a Great Circle. Small Circle. Zone. Spherical Segment. D. 14. B. VIII.

3. What are the *Three round bodies* treated of in the Elements of Geometry? D. 17. B. VIII.

4. The solidity of a Cone is equal to its base multiplied by a third of its altitude. P. 5. B. VIII.

5. Every section of a sphere, made by a plane, is a circle. P. VII. B. VIII.

6. The solidity of a sphere is equal to its surface multiplied by a third of its radius. P. XIV. B. VIII.

7. Demonstrate the above Propositions; also P. XVIII, and give General Scholiums. B. VIII.

BOOK IX.

1. Define a Spherical Triangle, Lune, Ungula, and the Pole of a Circle. D. B. IX.

2. "Two Symmetrical spherical triangles are equivalent." P. 16. B. IX.

3. "The sum of all the angles in any spherical triangle is less than six right angles, and greater than two." P. XIV. B. IX.

4. "The surface of a spherical triangle is equal to the excess of the sum of its three angles above two right angles multiplied by the tri-rectangular triangle. P. XVIII. B. IX.

Many other questions and Propositions might have been given, but the above will be sufficient to indicate what may be expected in regard to this branch.

CHAPTER XV.

NATURAL PHILOSOPHY.

REMARK.—The increased attention given to Philosophy in our schools at the present day renders it essential that *Teachers* should be well acquainted with this practical science. The references in the following questions on Natural Philosophy are to "Peck's Ganot."

NOTE.—The Author hesitated in making the selection of a Text-Book, to which to refer. But the *superior merits* of "Peck's Ganot" induced him to take that excellent work, trusting that as soon as it should become known, it would be in *general use.*

Art. stands for Article. Ex. for Example. Exp. for Experiment. P. for Page. Fig. for Figure.

INTRODUCTION.

1. What is Science? What is a Law? Define the Universe. P. 9.

2. Define Mind. Matter. In what two states may matter exist? P. 9.

3. What are the two divisions of science? P. 9.

4. What is Natural Philosophy? P. 9.

5. Into what may Natural Philosophy be divided? P. 10.

6. How may organized matter be divided? P. 10.

7. What are the corresponding divisions of General Physics? P. 10.

8. What is Physics Proper? Chemistry? P. 10.

9. What are the Pure Sciences? P. 10.

10. What are the Mixed Sciences? P. 10.

PRELIMINARY PRINCIPLES AND MECHANICS OF SOLIDS.

1. What are Physical Agents? Name them. Art. 1.

2. Define a Body. An Atom. A Molecule. Art. 2.

3. What are Molecular Forces? What is Attractive Forces? Repellent Forces? Art. 2.

4. What is the *Map* of a body? Density? How are bodies divided? Art. 3.

5. Define solids and fluids. How are fluids divided? Art. 4.

6. Define liquids, and gases or vapors. Art. 4.

7. What are the general properties of bodies? Art. 5.

8. Define Magnitude, Form, and Impenetrability. Art. 6, 7.

9. Define Inertia. Give examples of the principle of Inertia. Art. 8.

10. What is Porosity? Explain the porosity of gold by the Florentine. Exp. Art. 9.

11. Define a Filter. Divisibility. Compressibility. Dilatability. Art. 10–12.

12. What is Elasticity? Give examples of the most and least elastic bodies. Art. 13.

MECHANICAL PRINCIPLES.

1. Define Mechanics. When is a body at rest? In motion? Art. 14, 15.

2. Give examples of Rectilinear and Curvilinear Motion. Art. 16.

3. Give examples of Uniform Motion. Varied Motion. Define Velocity. Art. 17.

4. Define Accelerated and Retarded Motion. Art. 18.

5. Define and give examples of Forces, Powers and Resistances. Art. 19.

6. What three elements determine a force? Art. 20.

7. What is a Resultant of several forces? Illustrate. Art. 21.

8. Explain the parallelogram of forces. Art. 22.

9. Explain the flight of an Eagle. The sailing of a boat. Art. 23, 24.

10. What is the Resultant of parallel forces acting in the same direction? Art. 25.

11. Acting in different directions? Ex. Art. 25.

12. When are forces in equilibrium? Illustrate. Art. 26.

13. Define Centrifugal force. Centripetal force. Art. 27.

14. How does a body move when the centripetal force is destroyed? Art. 27.

15. State some of the effects of the centrifugal force. Art. 28.

16. What is a machine? A Motor? What is the advantage of machines? Art. 29.

17. Define a Lever. Tell how many classes. Example of each. Art. 30.

18. What are the *lever arms?* What is the relation between the power and resistance? Art. 31.

19. What is the relation between the power and velocity? Art. 31.

20. Is there any gain of power in using a lever? Ex. Art. 31.

21. Explain the scissors, and nut-cracker. Art. 32.

22. Name any other simple machines. Art. 33.

23. What machines are formed by combinations of simple machines? Art. 33.

24. What are the seven mechanical powers? Name them. Art. 33

PRINCIPLES OF GRAVITATION.

1. What is the force of Gravity? What is Universal Gravitation? Art. 34.

2. Explain the law of Universal Gravity. Art. 34.

3. State Newton's law. Art. 34.

4. State the effect of gravitation on the planets. Art. 35.

5. What are the orbits of planets? What is the force of Gravity? Art. 36.

6. What is the shape of the Earth? Define a vertical line. Art. 37.

7. Where do vertical lines meet? Illustrate by example. Art. 37.

8. When are vertical lines parallel? When not? Give example. Art. 37.

9. What are antipodes? Define a horizontal Line. Level. Art. 37.

10. What instruments are constructed on the principle of verticals and horizontals? Art. 37.

11. Define weight.   Center of Gravity.   Art. 40.

12. When is a body in equilibrium?   Give an example.   Art. 41.

13. What are the three cases of equilibrium?   Give example. Art. 42.

14. Define Stable, Unstable and Neutral equilibrium.   Art. 42.

15. What bodies are most stable?   Illustrate by example. Art. 43.

16. How do men and animals maintain a stable position?   Art. 43.

17. Where is the center of gravity in man?

18. Explain the principle of rope dancing.   Art. 43.

19. Define a balance.   Beam.   Fulcrum.   The scale.   Art. 44.

20. On what principle are bodies weighed?   Art. 44.

21. What are the requisitions for a good balance?   Art. 45.

22. State the methods of testing a Balance.   Art. 46.

23. What is the first law of falling bodies?   The second? Art. 48.

24. The third?   Illustrate each by an example.   Art. 48.

25. What is an Inclined Plane?   Explain its principle.   Art. 49.

26. How would you prove the third law of falling bodies? Art. 50.

27. What use is made of the inclined plane?   Art. 51.

28. What is a screw?   A wedge?   Art. 51.

29. What is a Pendulum?   Why does it vibrate?   Art. 52.

30. Explain the construction of the simple and compound pendulum.   Art. 53.

31. Explain the *laws* that *govern* the vibration of the pendulum.   Art. 54.

32. Mention some of the Applications of the Pendulum.   Art. 55.

33. Why do clocks lose time in summer, and gain in winter? Art. 55.

34. What is the length of a second pendulum in New York? Art. 55.

35. Describe a metronome. Art. 56.

PRINCIPLES OF MOLECULAR ACTION.

1. Define molecular forces. How divided? Art. 57.

2. Explain the effects of compressing and stretching bodies. Art. 57.

3. Define Cohesion. Adhesion. Give examples of each. Art. 58, 59.

4. Give example and explain the phenomena of Capillarity. Art. 60.

5. Give examples and illustrate the principles of Absorption and Imbibition. Art. 62, 63.

6. What principle is involved in the anecdote Pope Sixtus Quintus? Art. 63.

7. What effect will wetting ropes have on their length? Art. 63.

8. What is Tenacity? What bodies are most tenacious? Art. 64.

9. What is the form of greatest strength? Art. 64.

10. Define Hardness and Ductility. Malleability. Art. 65, 66, 67.

MECHANICS OF LIQUIDS.

1. Define Hydrostatics. Hydronamics. Art. 68, 69.

2. Say what you can concerning the properties of Liquids. Art. 69.

3. What is the principle of Pascal? Art. 70

4. State and explain the *law* in regard to the pressure of liquids. Art. 71.

5. How is the lateral pressure demonstrated? Art. 72.

6. Explain the upward pressure of liquids. Art. 72.

7. Explain the Hydrostatic Paradox. Art. 74.

8. What is the principle of the Hydraulic Press? Art. 76.

9. Illustrate the power of the Hydraulic Press by an example. Art. 76.

10. Explain the difference between equilibrium of solids and liquids. Art. 77.

11. What are the conditions of equilibrium in the case of liquids of different densities. Art. 80.

12. Describe a Water Level and its use. Art. 82.

13. Describe the principle of the spirit level. Art. 83.

14. Define a spring. Fountain. Artesian Wells. Art. 85.

15. Enunciate the principle of Archimides. Art. 86.

16. What is a Hydrostatic balance? Art. 87.

17. Explain the principles of Floating Bodies. Art. 89.

18. Give examples and illustrate the principles of Flotation. Art. 90.

19. Explain the action of the swimming bladder of Fishes. Art. 91.

20. What is the safest position in the water? Art. 92.

21. Define Specific Gravity. Art. 93.

22. How do you find the specific gravity of bodies? Art. 93.

23. What is Nicholson's Hydrometer? Art. 94.

24. How do you find the specific gravity of a liquid by the balance? Art. 95.

25. Which is the heaviest solid? Liquid? Art. 95.

26. Describe Beaume's Arcometer. Art. 96.

27. Describe the principle and object of the Alcoholometer. Art. 97.

28. Define the Lactometer and its use. Art. 98.

GENERAL PROPERTIES OF GASES AND VAPORS.

1. What is the difference between gases and vapors? Art. 99.

2. How many known gases are there? Which have not been liquified? Art. 99.

3. Describe the composition and uses of the Atmosphere. Art. 100.

4. How is the expansive force of air shown? Art. 101.

5. How can you show that *air* has weight? Art. 102.

6. What can you say of Atmospheric pressure? Art. 104.

7. How does pressure vary as we ascend? Art. 104.

8. Explain the *principle* of the Madgeburg Hemispheres, not the apparatus. Art. 106.

9. What is the pressure on the square inch? Art. 107.

10. Pascal's Experiment in detail, and his mode of reasoning. Art. 108.

11. Define a Barometer, and explain its principle. Art. 109.

12. Describe the Cistern Barometer in all its parts. The Thermometer. Art. 110.

13. Describe the siphon Barometer. Art. 111.

14. What are the requisites of a good Barometer? Art. 112.

15. Where are the fluctuations of the barometer greatest? Least? Art. 113.

16. How is the height of a Barometer for a day or year determined? Art. 113.

17. What are the causes of Barometrical fluctuation? Art. 114.

18. When does the barometer rise? Fall? Art. 114.

19. Explain the use of the Barometer as a weather glass. Art. 114.

20. Describe the difference between the Index and Siphon barometer. Art. 115.

21. On what principle can you measure the heights of mountains by a barometer? Art. 116.

22. What is height of the atmosphere? Art. 117.

23. How are pressures transmitted through gases? Art. 118.

24. What is the amount of pressure on the human body? Art. 119.

25. How is that pressure sustained? Art. 119.

26. Describe Mariotte's Law. Its consequence. His Tube. Art. 120, 121.

27. How is the *tube* used to verify *his law*? Art. 121.

28. Explain the Manometer and the different kinds. Art. 122.

29. Describe the Open Manometer. Closed Manometer. Art. 123, 124.

30. What is the object of the Manometers? Art. 124.

AIR PUMP.

1. What is an Air Pump? When and by whom invented? Art. 125.

2. Give a complete description of the air pump. Art. 125.

3. Explain clearly the action of the air pump. Art. 125.

4. How may the degree of rarefaction be measured? Art. 126.

5. Mention some experiments with the air pump. Art. 127.

6. How and why are articles of food preserved in vacuo? Art. 128.

7. What applications are made of this principle? Art. 128.

8. Explain the difference between the air pump and Condenser. Art. 129.

9. How is the degree of condensation measured? Art. 129.

10. State the effect of condensed air on combustion. Life. Divers. Art. 129.

11. Describe an Artificial Fountain. Art. 130.

12. Describe Hero's Fountain. How prepared for use? Art. 131.

13. Describe an Intermittent fountain. Art. 132.

14. Explain the principle of the Atmospheric Inkstand. Art. 133.

WATER PUMPS.

1. What is the difference between an air-pump and a water pump? Art. 125–134.

2. Describe in all its parts the Suction and Lifting pump. Art. 135.

3. Explain the action of this pump. Art. 135.

4. What and how many forms may be given to the force pump? Art. 136.

5. What is the difference between a Fire Engine and a pump? Art. 137.

6. How is the fire engine supplied with water? Art. 137.

7. How high may water be raised by the forcing pump? Art. 138.

8. Describe a Siphon, and tell when it may be used with advantage. Art. 139.

9. Explain the principle and action of the Siphon. Art. 139.

10. Describe the Siphon of constant flow. Art. 139.

BUOYANCY OF THE ATMOSPHERE.

1. Describe the principle of the Baroscope and its use. Art. 140.

2. Give the *law* of *buoyancy*, and tell when a body will rise in the air. Art. 140.

3. Describe a Balloon and its use. Art. 141.

4. What can you say of the history of ballooning? Art. 141.

5. With what are balloons filled? Tell how they are filled. Art. 142.

6. How is the ascensional power regulated? Art. 142.

7. What is the use of the barometer? Art. 142.

8. Describe a Parachute and its use. Art. 143.

9. Mention some remarkable balloon ascensions. Art. 144.

10. Describe the uses of balloons. The great American voyage. Art. 144.

A USTICS.

1. Define Acoustics. Sound. What is its cause? Art. 145, 146.

2. How is sound transmitted? What is a sonorous body? Art. 146.

3. What is a medium? Explain the vibrating cord. Art. 146.

4. How is sound imparted to the auditory nerve? Art. 148.

5. Explain how two sound waves produce silence. Art. 148.

6. What can you say of sound in vacuuo? Why? Art. 149.

7. What can you say of the propagation of sound in liquids and solids? Art. 150.

8. How is it shown that the earth transmits sound? Art. 151.

9. What is the velocity of sound? Art. 151.

10. Do all sounds travel with equal velocity? Art. 152.

11. Explain the reflection of sound. What is an echo? Art. 154.

12. Explain the causes of echoes. What is a Resonance? Art. 155.

13. State the causes that modify the intensity of sound. Art. 157.

14. How does wind modify sound? Art. 157.

15. What effect has a tube on sound? Art. 158.

16. Describe a Speaking Trumpet. Art. 159.

17. What is the difference between an Ear and speaking trumpet? Art. 160.

MUSICAL SOUNDS.

1. Define a Musical Sound. A noise. Art. 161.

2. What does Pitch depend upon? Art. 162.

3. What can you say of the limits of Audible Sounds? Art. 163.

4. What is a musical scale? Gamut? Why so called? Art. 164.

5. Define an Interval. A third. Fourth. An octave. An Accord. Art. 165.

6. Define a Consonance. Dissonance. Perfect accord. Art. 165.

7. Describe a tuning fork. Its use. Art. 166.

8. Of what are musical cords made? Art. 167.

9. State the *First* and *Second* laws governing vibrations. Art. 168.

10. Give the third and fourth laws governing vibrations. Art. 168.

11. How can you verify the preceding laws? Art. 169.

12. Describe a Sonometer and its use. Art. 169.

13. Upon what principles are stringed instruments made?

14. What can you say of sound from pipes? Art. 170.

15. Of pipes with fixed mouth-pieces? Art. 172.

16. Describe a Reed and some of the Reed instruments. Art. 173.

17. Describe the Bellows used with wind instruments. Art. 174.

18. Explain the different wind instruments. Art. 175.

HEAT.

1. Define Heat. Cold. Explain the theories of heat. Art. 177.

2. Describe the general effect of heat on solids. Art. 178.

3. How is the expansion in volume shown? Art. 179.

4. What can you say of the expansion of bodies by heat? Art. 179.

5. Is there any valuable use made of expansion and contraction? Art. 179.

6. Define sensible heat. Latent heat. Temperature. Art. 180.

7. On what principle is a thermometer constructed? Art. 181.

8. Describe the *best thermometer* in use. Art. 181.

9. Describe the process of making a thermometer. Art. 182.

10. Describe the mode of graduation. Art. 183.

11. Describe the three principal scales in use. Art. 184.

12. Explain the method of converting readings from one scale to another. Art. 185.

13. How does the alcohol differ from the mercurial thermometer? Art. 186.

14. Give some *Rules* for using thermometers. Art. 188.

15. Describe the two forms of Differential Thermometers. Art. 189.

16. Describe a Pyrometer and its principle. Art. 192.

RADIATION OF HEAT.

1. How does it appear that heat may be transmitted through space? Art. 193.

2. State and explain the Laws of Radiant heat. Art. 194.

3. Explain the mutual exchange of heat between bodies. Art. 195.

REFLECTION OF HEAT.

1. Define reflection of heat. Point of Incidence. Incident ray. Art. 196.

2. Define a reflected ray. What are the angles of *incidence* and *reflection?*

3. Give the *laws* which govern the reflection of heat. Art. 197.

4. Explain the principles of the concave mirror. Art: 198.

5. What can you say of the reflecting power of different substances? Art. 199.

6. Explain Leslie's method of determining the absorbing power of bodies. Art. 200.

7. Explain the Emission Power of a body. Art. 201.

8. State the *causes* which modify the reflecting power of bodies. Art. 202.

9. Illustrate the preceding principles by examples. Art. 203.

10. What can you say of the Conductibility of solid bodies? Art. 204.

11. State the principle in heating liquids. Art. 205.

12. Are liquids and gases good or bad conductors? Art. 206.

13. How are liquids and gases heated? Art. 205.

14. Give some applications of the preceding principles. Art. 207.

15. State the *laws* of expansion of solids, liquids and gases. Art. 208.

16. Give some examples illustrating the above laws. Art. 209.

17. Which is the most easily broken, a thick glass or a thin one? Why? Art. 209.

18. What effect has heat on a pendulum? Art. 210.

19. Explain the theory and construction of Harrison's Gridiron Pendulum. Art. 210.

20. Why are *liquids* more expansible than *solids?* Art. 211.

21. At what Temperature has water the greatest density? Art. 212.

22. What blessing flows from this exception to the general law? Art. 212.

23. State the law of expansion of gases. Art. 213.

24. Give some general examples in the applications. Art. 214.

25. On what does the density of a gas depend? Art. 215.

26. Define Fusion, and tell when it takes place. Art. 216.

27. Define latent heat. Sensible heat. Congelation. Art. 217, 218.

28. What is Crystalization? Give examples. Art. 219.

29. What is a freezing mixture? Give an example. Art. 220.

30. What is vaporization? Condensation? Give example. Art. 221.

31. Which is the most important vapor? Art. 221.

32. Describe volatile and fixed liquids. Art. 221.

33. What can you say of evaporation under pressure? Art. 222.

34. Why does a chestnut snap when roasted? Art. 222.

35. Why do vapors escape from the surface of liquids? Art. 223.

36. When does vapor cease to form? Example. Art. 224.

37. When is space saturated with vapor? Example. Art. 225.

38. State the causes that accelerate evaporation. Art. 226.

39. Define Ebullition. Give examples. Art. 227.

40. Give the causes that modify the boiling point of liquids. Art. 228.

41. What effect has the nature of the vessel on ebullition? Example. Art. 227.

42. Explain the principle of Papin's Digester. Art. 229.

43. What causes explosions of steam-boilers? Art. 229.

44. Explain Dalton's apparatus for measuring the tension of vapors. Art. 230.

45. What is latent heat of vaporization? Art. 231.

46. Give examples of cold produced by heat becoming latent. Art. 232.

47. Why does evaporation produce cold in surrounding bodies? Art. 233.

48. Explain the experiment with sulphurous acid. Art. 233.

49. Can mercury be frozen? How? Art. 233.

50. State what you can in regard to condensation of vapor. Art. 234.

51. What degree of heat is required for distillation? Art. 237.

52. Define an Alembic, and tell how distillation is effected? Art. 238.

53. How are gases liquefied? Art. 239.

54. Explain the apparatus and process of liquefying carbonic acid. Art. 239.

55. How do you compare bodies with respect to specific heat? Art. 240.

56. Define a Hygrometer. Illustrate. Art. 241.

57. Under what circumstances does the quantity of moisture in the air vary? Art. 242.

58. Explain the Hygroscope and its use. Art. 243.

59. Explain the principle of the hair Hygrometer. Art. 244.

60. What is meant by the hygrometric state of the atmosphere? Art. 245.

61. Explain how dew, fogs, clouds and rain are formed. Art. 246.

62. Define *frost, snow, hail* and *winds.* Art. 249, 250.

63. What are the causes of winds? Regular? Periodic? Variable? Art. 252.

64. Define an anemometer, and give the velocity of winds. Art. 254.

65. What are the sources of heat? Art. 255.

66. What are the causes of cold? Illustrate. Art. 256.

OPTICS.

1. Define optics. Sight. Explain the two theories of light. Art. 257–259.

2. What are the principal sources of light? Art. 260.

3. Define Opaque and Transparent bodies. Art. 261.

4. Explain the phenomenon of Absorption. Cause. Art. 262.

5. What is a ray of light? Pencil and Beam of light? Art. 263.

6. What is the velocity of light? By whom determined? Art. 264.

7. How does the intensity of light vary with the distance? Art. 265.

8. Describe the Photometer and its use. Art. 265.

9. What can you say of the Reflection of Light? Give examples. Art. 266.

10. Give and explain the laws of reflection. Art. 268.

11. What are mirrors? How are images formed by plane reflectors? Art. 272.

12. Explain the Nature of the images formed. Art. 273.

13. What is a curved mirror? Concave? Convex mirror? Art. 277.

14. Describe the principal focus of a concave mirror. Art. 278.

15. What are conjugate Foci? Radiant? Art. 279.

16. What can you say of the formation of images by concave reflectors? Art. 280.

17. Of the formation of images by convex reflectors? Art. 283.

REFRACTION.

1. Explain refraction and its cause. Art. 284.

2. Give and explain the laws of refraction. Art. 286.

3. Explain the method of proving refraction experimentally.

4. Explain some of the phenomena of refraction. Art. 289.

5. Why does a fish seem higher in water than he is? Art. 289.

6. Explain the phenomenon of total reflection. Art. 290.

7. On what principle do you explain the mirage? Art. 291.

8. Define and explain a prism. What effect has it on light? Art. 293.

9. Explain the course of a ray through a prism. Art. 294.

10. Which way is the ray bent? What is a Lens? Art. 295.

11. How many kinds, and of what are lenses made? Art. 296.

12. Name and describe the six kinds of lenses. Art. 297.

13. Where are the centres of curvature of a double convex lens? Art. 297.

14. Describe the principal Focus and focal distance. Art. 299.

15. Explain the action of a convex lens on sight. Art. 298.

16. Explain the conjugate foci and the laws. Art. 300.

17. Explain the formation of an image by a lens. Art. 301.

18. How does the size of the image compare with that of the object in different cases? Art. 301.

19. What kind of images are formed by convex lenses? Art. 302.

20. Explain the principle of Burning glasses. Art. 303.

21. Describe a Light-House. Art. 304.

22. Explain the reflectors used by Fresnel. Art. 304.

23. Define a Solar spectrum. Dispension. Art. 305.

24. Explain how a Prism acts to scatter rays. Art. 305.

25. Give the colors of the spectrum in their order. Art. 305.

26. What are heat and actinic rays? Art. 305.

27. What are simple colors? Why so called? Art. 306.

28. Explain how the rays of light may be reunited. Art. 308.

29. From what does the color of a body arise? Art. 309.

30. What is Newton's Theory of the colors of bodies? Art. 309.

31. What are complementary colors? What is the complement? Art. 311.

32. What is an accidental image? Explain the images of the Sun. Art. 312.

33. Explain the effect of looking at different colored cloths. Art. 312.

34. Explain the Rainbow. Primary and Secondary. How formed? Art. 313.

35. What is Chromatic aberration? Achromatic combinations? Art. 315.

OPTICAL INSTRUMENTS.

1. What are some of the most useful optical instruments? Art. 316.

2. What is a Telescope? How many classes are there? Art. 317.

3. What is the difference between the two classes? Art. 317.

4. Describe the Galilean Telescope. Explain the course of the rays in it. Art. 318.

5. Describe the Astronomical Telescope, and the course of the rays in it. Art. 319.

6. What is the difference between the Terrestrial and Astronomical Telescopes? Art. 320.

7. Describe the Reflecting Telescope and the Newtonian Telescope. Art. 322.

8. Describe Herschel's and Lord Ross's Telescope. Art. 323, 324.

1. What is a microscope? How constructed? Art. 325.

2. How is the magnifying power determined? Art. 326.

3. Explain the principle and construction of a compound microscope. Art. 327.

4. Describe the Magic Lantern, and method of using it. Art. 328.

5. How does the Phantasmagoria differ from the magic Lantern? Art. 329.

6. What is the polyrama? What is the Photo-Electric Microscope? Art. 331.

7. How is the circulation of the blood shown? Art. 331.

8. Describe the Solar Microscope. The Diorama. Art. 332.

9. Describe Camera Obscura. For what is it used? Art. 333.

10. How are the images made erect? Art. 334.

11. What is the most important application of the camera? Art. 336.

12. Give a sketch of the history and use of the Daguerreotype. Art. 336.

13. Explain the difference between Photography and Daguerreotyping. Art. 338.

STRUCTURE OF THE EYE.

1. Of what is the Eye composed? Use and optical properties? Art. 339.

2. Describe the eye in all its parts. The mechanism of vision. Art. 340.

3. What is the average limit of distinct vision? Art. 341.

4. Why is a person short-sighted? Long-sighted? Art. 341.

5. How are these defects remedied? Art. 342.

6. Explain how we are enabled to see clearly with two eyes. Art. 342.

7. Explain the theory and construction of the stereoscope. Art. 343.

MAGNETISM.

1. Define magnetism as a science. A magnet. Art. 345.

2. How many kinds of magnets are there? Describe each. Art. 345.

3. What remarkable property does the magnet possess? Art. 345.

4. Where is the strongest attraction? How shown? Art. 346.

5. What are poles? Equator? Give the Theory of magnets. Art. 347.

6. What other theory is there? Describe the magnetic action of the earth. Art. 347.

7. State the laws of attraction and repulsion. Art. 348.

8. What are magnetic bodies? Magnetized bodies? Art. 349.

9. Illustrate by examples. Explain the magnetic swan. Art. 349.

10. What is coercive force? How is it in different bodies? Art. 350.

DIRECTIVE FORCE OF MAGNETS.

1. What direction does a free magnet take? Art. 351.

2. How is a needle balanced? Is there any other way? Art. 351.

3. Why has the earth been regarded as a magnet? Art. 351.

4. Describe the magnetic meridian. Declination of the needle. Art. 352.

5. What is the annual variation? Perturbations? Illustrate. Art. 352.

6. What is a Compass? Describe it. What is its use? Art. 353.

7. How is the compass needle prevented from dipping? Art. 354.

8. What substances can be permanently magnetized? Art. 355.

9. How are bars of steel magnetized? Explain the method of single touch. Art. 356.

10. Describe a bundle of magnets. An Armature. Art. 357.

11. What are the advantages of the horse-shoe magnet? Art. 357.

ELECTRICITY.

1. Define Electricity as a science. Whence the name? Art. 358.

2. What can you say of the history of electrical discoveries? Art. 359.

3. What are the principal sources of electricity? Art. 360.

4. Describe the Electroscope. The Electrical Pendulum. Art. 361.

5. How is it shown that there are two kinds of electricity? Art. 362.

6. Explain fully the Hypothesis of the two Electrical fluids. Art. 363.

7. State the *laws* of electrical attraction and repulsion. Art. 364.

8. Describe Conductors and Insulators. Best conductors. Art. 365.

9. Give the method of electrifying bodies. Art. 366.

10. Where is the electricity of a body found? Art. 367.

11. What effect has the form of a body? Power of points? Art. 368.

12. Describe Induction. How is an insulated body affected by Induction? Art. 369.

13. Describe an Electrical Machine. When and by whom invented? Art. 370.

14. Explain the operation of the machine. Precautions in using it. Art. 373.

15. Describe the Electrophorous and its use. The Gold-leaf Electrometer and its use. Art. 375.

16. What is an electrical spark? How obtained? Art. 377.

17. How is a spark given? Describe an Electrical Stool. Art. 378.

18. Describe an Electrical chime. Electrical Puppet. Wheel. Art. 381.

19. Explain the Electrical Egg. Square. Cannon. Their use. Art. 384.

20. Describe the Electrical Condenser and its use. Art. 387.

21. Describe the *ways* in which a condenser may be discharged. Art. 388.

22. Describe the Discharger and its use. The Leyden Jar. Art. 390.

23. Describe the Electrical Battery. How charged and discharged. Art. 391.

24. Describe the condensing Electrometer and its use. Art. 393.

25. What are the physiological effects of Electricity? Art. 394.

26. What are some of the mechanical effects of Electricity? Art. 395.

27. Explain the heating power of Electricity. Art. 396.

28. Who first showed the identity of lightning and electricity? Art. 397.

29. What is Atmospheric electricity? Lightning? Thunder? Art. 399.

30. Describe the effects of the thunderbolt, and means of safety. Art. 402.

31. Describe Electrical meteors. Hail. Tornado. Art. 407.

32. Describe a Lightning-rod and the conditions that it should fulfill. Art. 404.

33. Describe the Aurora Borealis. What is its origin? Art. 408.

DYNAMICAL ELECTRICITY.

1. What is Galvanism? Why so called? What is Volta's theory? Art. 410.

2. Describe the Voltaic Pile. What are the poles? Electrodes? Art. 412.

3. What is the electric current? Chemical Theory of the Pile? Art. 414.

4. Describe the carbon Pile, and the effects of the Galvanic Battery. Art. 416.

5. How may a spark be obtained from a Battery? Art. 417.

6. Describe the heating and Illuminating effects of the Battery. Art. 419.

7. How is water decomposed by the Battery? Oxydes and Salts? Art. 422.

8. Explain the process of Electrotyping. Gilding and Plating. Art. 425.

ELECTRO-MAGNETISM.

1. What is the relation between magnetism and electricity? Art. 426.

2. Explain the action of the electrical current on the needle. Art. 427.

3. Give the principles established by Ampere and his Theory. Art. 430.

4. Describe the Galvanometer. Galvanic Multiplier, and its use. Art. 432.

5. How is an Iron bar converted into a magnet by magnetism? Art. 433.

6. Describe the Electro-magnet. The Electrical Telegraph. Art. 435.

7. Explain Morse's Registering Telegraph, and the mode of working it. Art. 436.

8. Explain Morse's Manipulator, and Receiver and Alphabet, and their uses. Art. 437.

9. What can you say of Induction by currents, and the properties of induced currents. Art. 441.

10. Describe the Physiological effects of electrical currents. Describe Electrical Fishes. Art. 443.

CHAPTER XVI.
CHEMISTRY.*

REMARK.—The Progressive spirit of the age has introduced Chemistry into our *Common Schools*. Hence candidates preparing themselves for *examinations* should not stop short of a *thorough* and *accurate* knowledge of this *noble Science*.

The references in the following questions on Chemistry are to "Wells' Principles of Chemistry." Art. stands for article. Ex. for example. Exp. for experiment. P. for page.

* The principal questions contained in the first four Chapters in Chemistry have been anticipated in the Chapter on Natural Philosophy.

1. What is Inorganic Chemistry? What is a Chemical Element? Art. 250.

2. Is any substance positively known to be elementary? Art. 250.

3. What is the number of Chemical Elements? Art. 251.

4. Into what two great classes are the elements usually divided? Art. 252.

5. How many of the elements are gaseous? How many liquid? Art. 252.

6. How are the elements distributed in nature? Art. 252.

7. In what condition are they found? Art. 252.

8. How are Compound bodies formed? Art. 252.

9. What is the difference between ancient and modern methods of explaining Chemical combination? Art. 254.

10. Define Affinity, and illustrate the characteristics of it. Art. 255.

11. Illustrate the third and fourth laws of Chemical Affinity. Art. 255.

12. Is the force of Affinity always the same? How shown? Art. 255.

13. Is matter under any circumstances ever destroyed? Art. 255.

14. Under what circumstances will combination occur? Art. 255.

15. Define Catalysis. Illustrate by an example. Art. 255.

16. What is understood by the Nascent state? Art. 255.

17. Do substances enter into combinations in all proportions? Art. 256.

18. State the *laws* which govern Chemical combinations. Art. 256.

19. State the *law* of definite proportions. Illustrate. Art. 257.

20. Explain the *law* of multiple proportions. Art. 258.

21. Demonstrate the *law* of equivalent proportions. Art. 259.

22. Explain the *law* of substitution. Chemical Equivalents. Art. 261.

23. Illustrate these *laws* by examples. Art. 260.

24. May the numbers expressing equivalents be varied and changed? Art. 261.

25. What is the unit of comparison in England and the United States? Art. 261.

26. Explain the *law* of combination by fixed equivalents. Art. 261.

27. Explain equivalent volumes. Art. 262.

28. On what Theory is Chemical combination explained? Art. 263.

29. How is the doctrine of equivalent proportions explained by the Atomic Theory? Art. 263.

30. Is there a relation between the atomic weight of an element and its capacity for heat? Art. 264.

CHEMICAL NOMENCLATURE AND SYMBOLS.

1. What three great classes of substances are recognized by Chemists? Art. 263.

2. Define an Acid. Give an example. What are Bases? Example. Art. 265.

3. Define an Alkali. Example. What are Salts? Example. Art. 265.

4. How may the properties of the Acids and Alkalies be illustrated? Art. 265.

5. What is the difference between an Acid and an Alkali? Art. 265.

6. What are Neutral bodies? Give example. Art. 266.

7. What is the origin of Chemical Nomenclature? Art. 267.

8. Explain the Nomenclature of the elements. Art. 268.

9. What are *binary compounds? Ternary?* Give example. Art. 269.

10. What are the compounds of Oxygen called? Chlorine, &c.? Art. 269.

11. What is a Protoxyd? Binoxyd? Give example. Art. 269.

12. How are Acid compounds of Oxygen named? Art. 269.

13. How are the different acid compounds distinguished? Art. 270.

14. How are Salts named? Give examples. Art. 271.

15. What do the prefixes hypo and hyper designate? Art. 271.

16. What two classes of Salts have been reorganized in Chemistry? Art. 271.

17. Why are symbols used in Chemistry? Art. 272.

18. Explain the symbols of Elements. Art. 273.

19. Explain the symbols of Compounds. Art. 273.

20. What are dernical formulæ? Art. 274.

21. How is the composition of Salts indicated by symbols? Art. 274.

22. What constituent is placed first? Art. 274.

23. Write out the proper symbols of Nitre.

24. What are Reactions and Reagents? Art. 275.

25. Explain Isomerism. Give example. Art. 276.

26. Explain Alstrophism. Give example. Art. 277.

27. Write out Marble by the proper symbols.

28. Write out Alcohol by the proper symbols.

29. Write out Chloroform by the proper symbols.

30. Spell Sugar, Chemically.

NON-METALLIC ELEMENTS.

1. How are the elements divided? Is this a natural division? Art. 278.

2. How many Metalloids are there? Name them. Art. 278.

3. What are the characteristics of the metalloids? Art. 278.

4. When and by whom was Oxygen discovered? Art. 279.

5. What can you say of the importance and distribution of Oxygen? Art. 280.

6. How is Oxygen usually procured? Give an example. Art. 281.

7. Describe the method of generating Oxygen from Chlorate of Potassa. Art. 281.

8. Why do you mix Manganese with the Chlorate of Potassa? Art. 281.

9. What is the chemical reaction in this experiment? Art. 281.

10. Do plants evolve oxygen? What experiment proves this? Art. 281.

11. What are the properties of oxygen? Illustrate. Art. 282.

12. Explain the action of oxygen on other substances.

13. What is spontaneous combustion?

14. Why do leaves, wood and fruit decay? Art. 282

15. What is a supporter of combustion? A combustible body? Art. 282.

16. Why do you usually apply heat to cause combustion to commence? Art. 282.

17. What analogy is there between combustion and respiration? Art. 283.

18. How may iron wire be burned? Explain. Art. 282.

19. What effect has pure oxygen on animal life? Art. 284.

20. Illustrate the various conditions under which oxygen exists in combination. Art. 285.

21. What is Ozone? By whom discovered? Art. 287.

22. How may Ozone be obtained? Art. 287.

23. How is Ozone proved to be simply modified oxygen? Art. 287.

24. What can you say of the daily consumption of oxygen? Art. 288.

25. What is said of the management of gases? Art. 289.

26. What precautions are necessary in collecting gases?

27. How may gases be transferred from one vessel to another? Art. 289.

28. Describe Gasometers. How constructed? Art. 290.

29. What effect will oxygen have on the *red* wick of a candle?

30. Explain iron rust and the burning of a candle.

HYDROGEN.

1. Give the history of Hydrogen, the symbol and density. Art. 291.

2. How is hydrogen obtained in the greatest purity? Art. 293.

3. Why does the Blacksmith sprinkle his fires with water? Art. 293.

4. What are the properties of hydrogen? Art. 294.

5. What is said of the lightness of hydrogen? Art. 294.

6. What is said of the inflammability of hydrogen? Art. 295.

7. Will oxygen and hydrogen unite of their own accord? Art. 296.

8. What are the best explosive mixtures of oxygen and hydrogen? Art. 296.

9. Explain the hydrogen gun. Art. 296.

10. What is said of the heating effects of the hydrogen flame? Art. 299.

11. Describe the oxyhydrogen blow pipe. Art. 300.

12. Describe the Drummond light. Art. 301.

13. What is said of the nature of hydrogen? Art. 302.

14. What compounds does hydrogen form with oxygen? Art. 303.

15. What is the composition of water? How formed? Art. 305.

16. Describe the Endiometer. Give the history of water. Art. 306.

17. What are the properties of water? Art. 307.

18. What can you say of the purity of water? Art. 310.

19. What can you say of spring waters? Mineral springs? Art. 312.

20. Of Saline Springs? Thermal Springs? River Water? Art. 314.

21. Why is the sea salt? Art. 316.

22. What can you say of the relative fitness of waters for use? Art. 317.

23. Describe hard water. Soft water.

24. How may the presence of air in water be demonstrated? Art. 320.

25. How may absolutely pure water be obtained? Art. 320.

26. What can you say of the solvent properties of water? Art. 321.

27. Of the chemical properties of water? Art. 322.

28. What is a hydrate? An anhydrous body? Art. 322.

29. Describe peroxide or binoxide of hydrogen. Art. 323.

30. How is it formed, and what are its properties? Art. 323.

NITROGEN.

1. What is the history of Nitrogen? Its nature and distribution? Art. 324.

2. What plants contain it in greatest abundance? Art. 325.

3. How is Nitrogen obtained? What are its properties? Art. 325.

4. What can you say of the combination of nitrogen? Art. 327.

5. What peculiarity has nitrogen in composition? Art. 328.

6. What is said of nitrogen in the animal system? Art. 328.

7. What is said of the elementary character of nitrogen? Art. 328.

THE ATMOSPHERE.

1. How was *air* regarded by the ancients? Art. 329.

2. When was the existence of separate gases first determined? Art. 329.

3. What is the composition of the air? Art. 330.

4. In what condition do oxygen and nitrogen exist in the air? Art. 330.

5. What is the proportion of Carbonic Acid in the air? Art. 330.

6. What is said of the uniformity of the condition of the atmosphere? Art. 330.

7. How much Ammonia is there in the air? Art. 330.

8. What office does Nitrogen appear to sustain in the atmosphere? Art. 330.

9. What is the magnetic condition of the atmosphere? Art. 330.

10. How is air analyzed? How are its elements determined? Art. 331.

11. Describe an Aspirator. Art. 331.

12. How may compounds of oxygen and nitrogen exist? Art. 332.

13. Describe Nitric Acid. Write the symbols for it. Art. 333.

14. Describe its history, distribution and preparation. Art. 334-336.

15. What circumstances led to its discovery? Art. 336.

16. Give the properties of Nitric Acid. What effect has light upon it? Art. 337.

17. Describe its chemical character. How does it act upon vegetable fibres? Art. 338.

18. Describe the action of Nitric acid on the metals. Art. 339.

19. Describe the composition of Nitrates. Art. 340.

20. What does N. O. stand for? Mention its nature. Art. 341.

21. How is protoxyd of Nitrogen prepared? Give its properties. Art. 343.

22. Describe NO_2. Name its properties. How prepared. Art. 345.

23. Name and describe NO_3.—NO_4.—NO_5. Art. 346, 347.

24. Is Hyponitrous Acid a vapor or liquid? Art. 346.

25. What are the properties of Peroxyd of Nitrogen? Art. 347.

CHLORINE.

1. Give the history of Chlorine. Distribution. Art. 348.

2. How is it prepared? Give its properties. Art. 350.

3. What precautions are to be observed in its preparations? Art. 350. ✦

4. What is the density of Chlorine? Can it be liquefied? Art. 351.

5. What combinations does chlorine form with water? Art. 351.

6. What are the relations of chlorine to combustion? Art. 352.

7. Why does phosphorus burn in chlorine with feeble light? Art. 352.

8. What effect does light have upon the mixture of chlorine and hydrogen? Art. 353.

9. What experiment illustrates this? Art. 353.

10. Explain the theory of bleaching by chlorine. Art. 354.

11. What is said of chlorine as a disinfecting agent? Art. 355.

12. Mention the compounds of Chlorine. Art. 356.

13. Describe H. Cl. Tell how prepared. Art. 357.

14. Explain the chemical reaction in this case.

15. Describe the Muriatic acid of commerce. Art. 360

16. What is Aqua Regia? How formed? Give its properties. Art. 361.

17. What is said of the oxyd of Chlorine? Art. 362.

18. What is hypochlorous acid? Art. 363.

19. Name and describe CAO. Cl. O., and give its properties. Art. 365.

20. Describe Cl. O_3, and give its properties. Art. 365.

21. Name and describe KO. Cl. O_5, and give its properties. Art. 367.

22. What is said of the Chloride of Nitrogen? Art. 369.

23. What was the original method of bleaching? Art. 370.

24. Describe the present method of bleaching. Art. 370.

25. What is the natural state of cotton fibres? Art. 370.

26. Give the whole process of bleaching cotton goods. Art. 370.

IODINE.

1. When and by whom was Iodine discovered? Art. 370.

2. Give its natural history and distribution in nature. Art. 371.

3. How is it formed? Give its properties. Art. 373.

4. What effect has Iodine on the metals? Art. 374.

5. What is the test of Iodine? Describe its principal salt. Art. 374.

BROMINE.

1. Who discovered Bromine? How obtained? Properties? Art. 375.

2. How does Bromine act upon the metals? Art. 378.

3. What are its uses and compounds? Art. 378.

FLUORINE.

1. Describe Fluorine. Why is it difficult to obtain it? Art. 379.

2. Describe Hydrofluoric acid. Tell how prepared. Art. 379.

3. Describe the following compounds: CAF. SO_3. HO. CaO. SO_3. HF.

SULPHUR.

1. What is the history of Sulphur? Art. 380.

2. Describe its properties. What are its commercial forms? Art. 381.

3. What is its affinity for other elements? Art. 382.

4. In what two forms does Sulphur crystalize?

5. What is the Milk of Sulphur? Art. 384.

6. Describe the compounds of Sulphur and oxygen. Art. 385.

7. Describe SO_2, its production and properties. Art. 387.

8. Describe how SO_3 is made on a large scale. Art. 388.

9. What is Nordhausen Sulphuric Acid? Art. 390.

10. Explain the action of SO_3 on the metals and fibres. Art. 392, 332.

11. Describe SO_2 and HS. Give their properties. Art. 395.

12. Why do surfaces painted with lead blacken on exposure to this gas? Art. 395.

13. Why are zinc paints, for many situations, preferable to lead? Art. 395.

14. What effect has tellurium upon the animal system? Art. 397.

PHOSPHORUS.

1. Give the history of Phosphorus and its distribution. Art. 398.

2. How is phosphorus obtained? Give its properties. Art. 401.

3. What is said of its solubility and inflammability? Art. 401.

4. Why is phosphorus good for the manufacture of matches? Art. 401.

5. What experiments illustrate the characteristics of phosphorus? Art. 401.

6. Describe the poisonous properties of phosphorus. Art. 401.

7. Describe Allotropic or Atmospheric phosphorus. Art. 402.

8. Relate the history and origin of matches. Art. 403.

9. How is phosphoric acid prepared? Art. 405.

10. What are its properties? Art. 405.

11. Describe Phosphoretted Hydrogen, and tell how it is prepared. Art. 407.

12. What phenomenon attends its evolution in air? Art. 407.

13. Describe its properties. Describe the Will O' the Wisp. Art. 407.

BORON.

1. What is said of Boron? Its properties? Art. 410.

2. Describe Boracic acid. How is it collected? Art. 410.

3. What is a flux? Why is borax valuable as a flux? Art. 411.

SILICON.

1. Relate the history of Silicon. Is the pure element known? Art. 412.

2. Describe Silica. Quartz. Amethyst. Sand. Art. 413.

3. Give the properties of each. Art. 414.

4. What are examples of natural silicates? Art. 414.

5. Describe fluosilicre acid. Art. 415.

CARBON.

1. What can you say of Carbon? Its specific gravity? Art. 416.

2. In what condition is carbon found naturally? Art. 416.

3. Under what circumstances is the diamond found? Art. 417.

4. How is it cut? In what form is it cut for jewelry? Art. 417.

5. What is the origin of the diamond? Art. 417.

6. How large a diamond has ever been found? Art. 418.

7. Have any attempts been made to manufacture diamonds?

8. What is a graphite? What is said of its use? Art. 419.

9. Describe gas, carbon, mineral coal, its properties. Art. 421.

10. Describe Anthracite coal. Coke. Charcoal. Art. 425.

11. How is Charcoal formed? Soot? Lampblack? Properties of each? Art. 427.

12. What is said of the compounds of carbon and oxygen? Art. 428.

13 Describe CO_2. Preparation and properties. How solidified? Art. 433.

14. What are the properties of solidified gas? Art. 433.

15. What are petrifactions? Art. 435.

16. Describe the natural production of CO_2. Art. 436, 437.

17. Explain the formation of the carbonates. Art. 438.

18. Describe CO. formation and properties. Art. 439.

19. What is Cyanogin? How prepared? Properties?' Art. 442.

20. Give the analysis of Prussiate of Potash and Prussian Blue. Art. 444.

21. Give the analysis of Prussic Acid and properties. Art. 447.

22. How is it supposed to occasion death? Art. 447.

23. Describe organic acid. Art. 450.

24. What is light Carburetted hydrogen? Art. 452.

25. Describe C_4. H_4. preparation and properties. Art. 454.

26. How is illuminating gas made? Coal gas? Art. 456.

27. Describe the whole process of making and purifying coal gas. Art. 457, 458.

28. Describe the explosive compounds of illuminating gas. Art. 459.

29. Give the history of the introduction of gas. Art. 460.

COMBUSTION.

1. What was the original supposition concerning fire? Art. 462.

2. Explain the general principles of the phlogistic theory. Art. 462.

3. By what experiment was the phlogistic theory overturned? Art. 463.

4. Define Combustion. Supporters of combustion. Art. 464.

5. Define Combustibles and Burnt bodies. Art. 465.

6. Describe the difference between combustion and explosion. Art. 466.

7. What is the origin of heat in combustion? Art. 467.

8. Is the quantity of heat increased by the rapidity of combustion? Art. 469.

9. Is any matter lost during combustion? Art. 470.

10. What are the ordinary products of combustion? Art. 472.

11. What good does blowing the fire do? Art. 473.

12. How much heat will a pound of charcoal in burning evolve? Art. 474.

13. Upon what does the light which accompanies combustion depend? Art. 475.

14. Describe flame. Art. 475.

15. What are the materials for illumination? Art. 476.

16. Explain the combustion of a candle. Art. 477.

17. Explain the structure of flame. Art. 478.

18. What is essential to the existence of flame? Art. 480.

19. Explain the *principle* of Davey's safety lamp,—not the lamp. Art. 481.

20. Give the requisites for an artificial light. Art. 482.

21. Describe the Argand lamp. Art. 483.

22. Describe the Blow Pipe. Art. 483.

METALIC ELEMENTS.

1. What can you say of the abundance of the metals? Art. 487.

2. What are the essential characteristics of the metals? Art. 488.

3. What is said of their hardness? Density? Malleability? Ductility? Art. 488.

4. Of their *Tenacity? Fusibility? Welding?* Volubility? Art. 488.

5. Describe Alloys. Amalgam. Do all the metals crystalize? Art. 491.

6. Describe the metals in the order of their affinity for oxygen. Art. 492.

7. How may the metals be classified? Art. 493.

8. Which are the Noble metals? Why so called? Art. 493.

9. Describe Potassium. Give its symbol. Specific gravity. Art. 494.

10. What is said of its distribution? Preparation? Properties? Art. 497.

11. Describe KO. How prepared? Properties? Caustic Potassa? Art. 499.

12. Describe KO. CO_2. KO_2. CO_2. Give the properties of each. Art. 502.

13. Describe Niter. Give the symbols. Properties. Art. 503, 504.

14. Describe Gun-powder. Symbols, properties and elements. Art. 505.

15. How is Gun-powder manufactured? Art. 506.

16. Is the explosion of Gun-powder instantaneous? Art. 506.

17. How is the goodness of powder tested? Art. 506.

SODIUM.

1. Describe Sodium. Where does it occur in nature? Art. 506.

2. Describe NaO. HO. Also Na Cl., and give the properties of each. Art. 510.

3. Is common Salt a Chemical salt? Why? Art. 512.

4. What proportion of salt exists in sea water? Art. 512.

5. What can you say of the solubility of salt? Art. 512.

6. Say what you can of NaO, SO_3+10HO. Art. 513.

7. Describe in detail NaO, CO_2+10HO. Art. 514.

8. Describe a reverberatory furnace. Art. 514, Note.

9. Give the history and introduction of Carbonate of Soda. Art. 514.

10. Describe Bicarbonate of Soda. Also NaO. NO_5. Art. 515, 517.

AMMONIUM.

1. What is Ammonium? Give its symbol. Art. 519.

2. Analyze Sal-Ammoniac, and give its symbols. Art. 520.

3. Describe N. H_4O. The preparation and properties. For what used? Art. 521.

4. Analyze NH_4S+HS. What are the properties of the Alkalies? Art. 528.

5. Which are the Alkaline earths? Art. 529.

BARIUM AND STRONTIUM.

1. Describe Barium and BaO, and BaCl. Art. 530.

2. Describe Strontium. Also SrO. For what used? Art. 531.

CALCIUM.

1. What is Calcium? Its equivalent? Symbol? Art. 532.

2. Give the properties and atomic weight of CaO. Its use. Art. 533.

3. What can you say of Mortars and Cements? Art. 535.

4. Describe Hydraulic Cements. CaO. CO_2. Art. 536, 537.

5. How may the durability of stone be tested? Art. 539.

6. Describe in detail CaO, $SO_3 + 2HO$; also Ca, Cl. Art. 540-542.

MAGNESIUM.

1. Describe Magnesium. Where found? Art. 543.

2. Analyze Mgo. Give the symbols and properties of Epsom Salts. Art. 545.

3. Also of Mgo, CO_2. Give the properties of the alkaline earths. Art. 547.

ALUMINIUM.

1. Describe Aluminium and its Oxyd. Art. 549, 550.

2. Describe and name Al_2O_3. $3SO_3 + KO$, $SO_3 + 24HO$. Art. 551.

3. How is Alum manufactured? Art. 552.

4. How does Alumina act in dyeing? What are lakes? Art. 552.

5. What is Carmine? Clay? Art. 553, 555.

GLASS AND POTTERY.

1. What is Glass? How made? Bohemian glass? Art. 557, 558.

2. What is the composition of Common white glass? Art. 558.

3. Analyze Green Glass. Flint Glass. Colored glass. Art. 559.

4. Describe Enamel. Annealing. Art. 561, 562.

5. Of what is porcelain made?   For what is it used?   Art. 564.

IRON.

1. Give the history and distribution of *Iron*.   Art. 565.

2. Is malleable iron found in nature?   How made?   Art. 565.

3. How may chemically pure iron be obtained?   Art. 565.

4. Name and describe the compounds of iron and oxygen. Art. 566.

5. What are the principal ores of iron?   Art. 571.

6. Name and describe Fe S_2.[1]   For what is iron used?

7. Describe Cast Iron.   The melting of Iron.   Art. 577.

8. Describe bar-iron.   Steel.   Give the properties of each. Art. 579.

MANGANESE AND CHROMIUM.

1. What are the properties of Manganese?   Where is it found? Art. 582.

2. Mention the compounds of Manganese.   Art. 582.

3. Describe Chromium and all its compounds.   Art. 583.

4. Name PbO. CrO_3.   Also CaO_3.   Art. 584, 585.

5. For what are the last two substances used?   Art. 585.

COBALT AND NICKEL.

1. Describe Cobalt and its oxyd.   Also Sympathetic Ink. Art. 586, 587.

2. What is Nickel?   Analyze German Silver.   Art. 588.

ZINC AND CADMIUM.

1. Describe Zinc.   Its properties.   Its equivalent.   Art. 590.

2. How is Zinc reduced from its ores?   Art. 590.

3. What is galvanized iron?   What is ZnO?   For what uses? Art. 593.

4. Describe Cadmium.   Art. 594.

LEAD AND TIN.

1. What is said of the distribution of *lead?*   What is Galena? Art. 595.

2. What are the properties of lead?   Its compounds?   Art. 596.

3. What effect does water have on lead? Art. 597.

4. What salts arrest the action of water on lead? Art. 597.

5. Describe and name PbO.—2PBO. PbO$_2$—PbO. CO$_2$. Art. 598, 599.

6. What are antidotes to lead poisoning? Art. 600.

7. How are shot made? Composition? Art. 601.

8. What is Zinc? The *cry* of *tin*? What is putty powder? Art. 603.

9. Describe Sn Cl. and Sn S$_2$. What is tin plate? Art. 605.

10. Analyze Britannia metal and Pewter. How are pins made? Art. 605.

COPPER AND BISMUTH.

1. What is said of the occurrence of Copper in Nature? Art. 606.

2. What are the properties of Copper? Why does it corrode? Art. 607.

3. What is the best solvent of Copper? Art. 608.

4. Describe and give the properties of CuO—CuO$_2$—CuO, SO$_3$. Art. 610, 611.

5. Analyze CuO, NO$_5$. Also Verdigris. Art. 612, 613.

6. Describe the characteristics of the salts of copper. Art. 614.

7. Analyze brass. Gun-metal. Bell-metal. Bronze. Art. 615.

8. Describe Bismuth and its compounds. Art. 616.

ANTIMONY AND ARSENIC.

1. What can you say of Antimony, its use and properties? Art. 618.

2. What are the chief compounds of Antimony? Art. 619.

3. Describe Tartar Emetic. For what used? Art. 620.

4. In what form does Arsenic occur in Nature? Art. 621.

5. How is the Arsenic of commerce prepared? Art. 621.

6. Describe AsO$_3$ and AsO$_5$. . Give their properties. Art. 622.

7. Give the tests for Arsenic in detail. Art. 625, 626.

8. How may Antimony be distinguished from Arsenic? Art. 626.

9. What amount of Arsenic is fatal? Art. 626.

MERCURY.

1. What can you say of Mercury? Art. 628.

2. At what temperature does mercury freeze? Boil? Art. 629.

3. What is said of its power to resist oxydation? Art. 629.

4. Describe Blue Pill. Mercurial Ointment. Art. 631.

5. Analyze Calomel and $HyCl.$—Hy_2O. Art. 632–635.

6. Describe Hy S. What are the uses of Mercury? Art. 637, 638.

7. Describe the alloys of Mercury. How are mirrors made? Art. 639.

SILVER.

1. Give the history of silver, and its distribution. Art. 640.

2. How is silver obtained from the ores? Art. 641.

3. How is silver freed from lead? What is a Cupel? Art. 643.

4. What are the solvents of silver?

5. Describe Ago NO_5. For what is it used? Art. 647.

6. Analyze Ag Cl. Give its uses. Art. 649.

7. What is standard silver in U. S. and Great Britain? Art. 649.

8. How may articles be silvered? Plated? Art. 649.

9. What is dead silver? What is a test for silver? Art. 649.

10. How may glass be silvered? Art. 650.

GOLD.

1. Give the Natural history of Gold. Art. 651.

2. How is gold obtained from the earth? Art. 651.

3. What are the properties of gold and its compounds? Art. 652.

4. How is perchloride of gold prepared? For what used? Art. 653.

5. In what condition is gold used in the arts? Art. 654.

6. How is the purity expressed? How obtained? Art. 654.
7. What is Assaying? Quartation? Bullion? Art. 655.
8. How is fine gold prepared? Gold leaf? Art. 656.

PLATINUM.

1. How and where is platinum formed in nature? Art. 657.
2. Mention the properties of platinum. Its infusibility. Art. 658.
3. What can you say of its solubility? Uses? How manufactured? Art. 658.
4. Describe PtO—PtO_2—$PtCl$—$PtCl_2$. Art. 659.
5. How is spongy platinum prepared? Platinum Black? Art. 659.

ORGANIC CHEMISTRY.

1. Define Organic Chemistry.
2. Describe the composition of organic substances. Art. 666.
3. How are so many different organic compounds produced from so few elements? Art. 666.
4. What organic bodies as a class are generally wanting in Nitrogen? Art. 666.
5. What are characteristics of organic and inorganic bodies? Art. 667.
6. What circumstances attend the decomposition of organic bodies? Art. 667.
7. What is the principal origin of organic substances? Art. 668.
8. What are Compound Radicals? Describe them. Art. 669.
9. Describe the essential immediate principals of plants. Art. 670.
10. Describe Vegetable Tissue. Starch. Gum. Sugar. Art. 671.
11. Name and describe $C_{12} H_{10} O_{10}$. Analyze Gum. Cotton. Art. 674, 675.
12. Describe Collodian, Parchment, Paper. Art. 676–678.
13. What can you say of the destructive distillation of wood? Art. 679.

14. Of Pyroligneous Acid? Creosote? Tar? Mineral Oils? Art. 680–683.

15. Of Asphaltum? Of the contents of the cells of plants? Art. 684, 685.

16. What is the difference between *wood*, sugar and gum? Art. 690–693.

17. What is the difference between Cane and Grape Sugar? Art. 702.

18. Analyze and describe *Albumen*, Caseine and Gluten. Art. 704.

19. What is the difference between Proteine and Albuminous substances? Art. 708, 709.

20. What distinguishes living from dead organized matter? Art. 711.

21. What is putrefaction? Fermentation? Yeast? Art. 713–715.

22. Describe the different kinds of fermentation. Art. 716.

23. What can you say of poisons, contagions? Art. 719.

24. Name and describe $O_2 \; C_4 \; H_6$, and all its derivatives. Art. 721.

25. Analyze Beer, Wine, Ardent Spirits, Bread. Art. 722–729.

26. Describe all the *products* of the action of *acids* upon alcohol. Art. 732.

27. Analyze Ether, Chloroform, and point out the difference. Art. 741.

28. Analyze and describe *all* the vegetable *acids*. Art. 745.

29. Name the properties of $C_2 \; O_4 \; H_6 - C_8 \; H_6 \; Ok_2$—Tannin. Art. 746, 747.

30. How is Leather made? Ink? Gallic Acid? Art. 751–753.

31. Describe the Organic Alkalies. Also $C_{35} \; H_{21} \; NO_7$. Art. 755.

32. Describe and give the properties of Quinine and Strychnine. Art. 757, 758.

33. Describe the Organic Coloring substances. Calico printing. Art. 761–764.

34. Say what you can of Oils, Fats and Resins. Art. 769.

35. What can you say of Elastic Gums? Gutta Percha? Art. 779–795.

36. Describe briefly the nutrition and growth of plants, soils and manures. Art. 796–804.

37. What can you say of animal organizations? Art. 807.

38. Describe the proximate animal constituents. Art. 808.

39. What can you say of the composition of the Brain? Nerves? Skin? Art. 814, 815.

40. What is the composition of Hair? Wool? Hoofs? Teeth? Art. 816–818.

41. Describe the composition of the Blood. Its circulation. Art. 824.

42. Describe digestion and its functions. Nutrition. Art. 826.

43. Describe in detail *respiration* and the composition of the lungs. Art. 826.

44. What are the uses of respiration? Is the skin a respiratory organ? Art. 827.

45. Describe the nature and functions of food. Art. 828.

46. What are the relative values of different meats and vegetables as nutritive qualities? Art. 831.

47. What can you say of the relation between animals and plants? Art. 832.

CHAPTER XVII.
SCIENCE OF GOVERNMENT.

REMARK.—In these days of *National trial*, every person permitted to *cross the threshold* of the *school-room*, as a *teacher* of youth, should be thoroughly imbued with the *fundamental principles* of his Government, and *breathe nought* but the *vital air* of *pure loyalty*.

NOTE.—The references in the following questions on the SCIENCE OF GOVERNMENT, are to MANSFIELD'S POLITICAL ECONOMY. Art. stands for article. P. for page.

1. What do you understand by *Sovereignty?* Art. 1.

2. Give clearly your idea of *Government.* Art. 2.

3. Define Law. What is a Constitution? Art. 3, 4.

4. Define a Despotism. A Monarchy. Art. 5, 6.

5. Explain clearly the difference, if any, between a Republic and a Democracy. Art. 7, 8.

6. In what does a Republic or Democracy differ from an Aristocracy? Art. 7, 8, 9.

7. Wherein did the Republic of Athens differ from that of the "United States?" Art. 7.

8. What is the difference between a Party and a Faction? Art. 10, 11.

9. What is a Legislature? Congress? Art. 12, 13.

10. How does Legislative power differ from Executive? Art. 14, 15.

11. Describe the Judiciary. Art. 16.

12. In what does *Statute* Law differ from *Common Law?* Art. 17, 18.

13. Which is superior, *Common* or Statute Law? Art. 18.

14. What is a Corporation? A Charter? A Court? Art. 19, 20.

15. Define the terms, *Municipal.* Jurisdiction. Art. 22, 23.

16. What do you understand by Impeachment? Crime? Art. 24–27.

17. Explain the difference between a Verdict and a Judgment. Art. 25, 26.

18. What is TREASON? Explain in *detail* in what it consists. Art. 28.

19. What is the difference between TREASON and Revolution? Art. 28–32.

20. Define Felony. Reprieve. Diplomacy. Art. 29, 30, 31.

21. What is the distinction between Diplomacy and Aristocracy? Art. 31.

22. What is an *Ex Post Facto Law?* Art. 33.

23. Describe a Bill of Attainder. Art. 34.

24. What do you understand by a Bill? Revenue? Art. 35, 36.

25. What is a *Treaty?* Art. 37.

26. In what way does a Foreigner become a citizen? Art. 38.

27. What do you understand by Bankruptcy? Art. 39.

28. What is a Test Act?   Art. 40.

29. Define the terms *Ballot* and Quorum.   Art. 41, 42.

30. What do you understand by Majority and Minority?   Art. 43.

31. What is the distinction between a Majority and Plurality?   Art. 44.

32. What is an Indictment?   A Grand Jury?   Art. 45, 46.

33. What do you understand by *Taxes?* A Legal Tender?   Art. 47, 48.

34. How many forms of Government were there originally in the Colonies?   Art. 2.   P. 29.

35. Describe the *Proprietory Government.*   Art. 4.

36. Describe the *Charter Government.*   Art. 3.

37. In what respect did the Royal Government differ from the Charter or Proprietory Government?   Art. 3, 4, 5.

38. What can you say in regard to the Articles of Confederation?   Art. 17.

39. For what did the *Articles* of Confederation provide?   Art. 23.   P. 37.

40. What were some of the Obvious deficiencies in the Articles of the Confederacy?   Art. 25.   P. 40.

41. In what did the idea of UNION and the CONSTITUTION of the United States originate?   Art. 27.

42. What is asserted in the PREAMBLE of the Constitution?   Art. 29.   P. 42.

43. What are the *objects* proposed in the Constitution?   Art. 32.

44. How many Articles does the *Constitution* of the U. S. contain?   Art. 33.

45. How many *Amendments* have been made to it?   Art. 33.

46. To what does the *first Article relate?*   Art. 33.

47. To what do the several *Articles relate?*   Art. 33.

48. What rights are guaranteed in the First Amendment?   Art. 433.

49. What *principles* are *modified,* or *rights secured* in each of the Amendments?   Art. 432, 442.   P. 203, 207.

50. By whom was the Constitution proposed?   Art. 31.   P. 43.

51. Who ratified the Constitution or gave it power? Art. 31.

52. Explain clearly how the members of Congress are elected. Art. 37–50.

53. Explain in detail how the President of the U. S. is elected. Art. 290.

54. What are the duties of the President? Art. 308. P. 145.

55. How is the Vice President chosen? Art. 290. P. 136.

56. What are the duties of the Vice President? Art. 304.

57. What can you say of the Cabinet of the U. S.? Art. 520.

58. What are the duties of the Secretary of State? Art. 520.

59. What are the duties of the Secretary of War? Art. 539.

60. What can you say of the Department of the Treasury? Art. 538.

61. Describe the general duties of the Navy Department. Art. 541.

62. Give the general duties of the Post Office Department. Art. 543.

63. Mention some of the duties of the Home Department. Art. 544.

64. How many standing committees are there in the House of Representatives? Art. 561.

65. Name the committees. Art. 561.

66. What can you say of the theory of State Government? Art. 476.

67. What can you say of the practical operation of the State Government? Art. 578.

68. What is the difference between the Government of the U. S. and that of Great Britain? Art. 6–9. P. 16, 17.

69. How does the mode of electing a Governor differ from that of electing the President of the U. S.? Art. 290.

70. How does the mode of electing a Representative to Congress differ from that of electing a Senator to Congress? Art. 37–54. P. 45–52.

71. How many votes is each State entitled to cast for President? Art. 492.

72. What is the difference between the National and State Courts? Art. 583.

73. In what way are the Judges of the National and State Courts appointed to office? Art. 311.

74. What can you say of the Writ of Habeas Corpus? Art. 231–600.

75. Upon what does the Government of the U. S. rest? Art. 502.

76. In what way can you best increase the *Love* of Country in every child in our Schools?

CHAPTER XVIII.

MUSIC.

In many schools a part of the time of every pupil is given to Music. More time would be given if the teachers were competent to give instruction in this *much neglected* but very important branch. The demand for "Teachers who can sing" is increasing. Not long since one school rejected twenty-seven teachers simply because they could not lead the school in Vocal Music,—yet who otherwise stood high in their profession. All teachers should understand the principles of music; it would please and profit them.

> "Let the people praise Thee, O God,
> Yea, let all the people praise Thee."

1. What is Music?

2. What can you say of a Musical tone?

3. How many essential properties has such a tone?

4. Name these properties or characteristics.

5. Into how many departments is music divided?

6. What can you say of *Rythmics? Melodies? Dynamics?*

7. How is length of tone indicated to the eye?

8. How are measures expressed?

9. Describe *Double Measure.*

10. What do you understand by Beating Time? Its object?

11. What is Accent?

12. What are signs of Tones? Signs of Silence?

13. What is the difference between the office of a *single bar* and a *double bar?*

14. What do you understand by the scale?

15. Of how many *tones* does the scale consist?

16. What are the names of the tones of the scale?

17. Can you give those tones accurately? Make them.

18. What is the object of the Staff?

19. What are Notes? What are degrees of the Staff?

20. What is the office of the spaces and added lines of the Staff?

21. What syllables represent the tones?

22. How are the syllables pronounced?

23. How is the Scale represented on the Staff?

24. What do you understand by Absolute Pitch Letters

25. Have the letters a fixed position on the scale?

26. What do you understand by Key of C?

27. What is the Clef? Clef-letters?

28. What can you say of the F Clef? G Clef?

29. What can you say of prolonged tones?

30. What is the Primitive form of measure?

31. How does primitive form of measure differ from derived form?

32. Describe Triple Measure.

33. Where does the *Accent occur* in Triple Measure?

34. What is the distinction between double and triple time?

35. What characters are used to designate the kind of measure?

36. Describe Quadruple Measure.

37. Where does the Accent occur in Quadruple Measure?

38. Name the different kinds of Notes and Rests.

39. What is a syncopated tone? Is it accented or not?

40. What are skips, and between what tones do they occur?

41. To what tone does the tone seven naturally lead?

42. To what does four naturally lead?

43. What can you say of the Extension of the Scale?

44. Into how many classes is the human voice naturally divided?

45. What is the distinction between the Bass and Tenor?

46. What is the difference between the Alto and Treble?

47. How do you distinguish between the different tones denoted by the same letter?

48. How many octaves are embraced in the whole compass of tones appreciable by the human ear?

49. How many of these octaves are within the range of the human voice?

50. What is an Interval? What are Steps? Half Steps?

51. Where do the whole steps and half steps occur?

52. Describe Sextriple measure. Compound measure.

53. Describe the *Minor Scale*. The Natural Minor Scale.

54. What are Triplets? What is a Chromatic Scale?

55. What is a Sharp or Flat? How far do they continue?

56. What is the office of D Natural, as used in Music?

57. Explain the following, viz.: *Every Major Scale* has its *Relative Minor*, and *every Minor Scale* has its *Relative Major*.

58. What do you understand by the *Transposition* of the Scale?

59. What do you understand by the Key Letter?

60. What must be preserved in the Transposition of the Scale?

61. Explain the Transposition of the Scale by Fifths.

62. What is the Signature of the key, and where is it placed?

63. What can you say of the Relations of Tones? Tone of Transposition?

64. Explain the *transposition* of the scale by *fourths*.

65. What is a Trill? Turn?

66. Define Mezzo. Piano Forte. Pianissimo. Fortissimo. Legato. Staccato.

67. What do you understand by *Swell? Crescendo?* Diminuendo? How are they indicated?

68. What do you understand by the *purity* of *tone?*

69. What are *Tonic Elements?* What are common errors in singing?

70. What *reasons* can you give *why music should be taught in all our schools?*

CHAPTER XIX.

THEORY AND PRACTICE.

REMARK.—The *principles* implied in the *foregoing questions* will be of but little avail to *Candidates*, unless they have a good *theory* of *imparting instruction;* a theory which they can practice. One that is *simple, pliant,* natural, as opposed to one that is complex and difficult.

The following questions are therefore proposed in order that candidates may have an occasion to draw upon their own resources, and thereby see whether they have a clear system in mind, or whether they must go before their schools to experiment. Such should remember that the *material* upon which they are to work is *too expensive* for experiments.

The answers to many of the following questions may be found in "Page's Theory and Practice," and "Northend's Teacher and Parent." Yet many of them are *unwritten*, hence must be looked for in the *mind*. This is as it should be, as in this department no *superior teacher* is a *mere imitator*.

1. How would you organize a school?

2. What spirit should one possess who has access to the *sanctuary* of the mind?

3. What motive should govern the teacher, moral or pecuniary?

4. What especial preparation does he need who is to play upon the "harp whose tones, *whose living tones,* are left forever in the strings?"

5. What can you say of the responsibility of the *teacher?*

6. What can you say of the ventilation of the school-room?

7. What should be the appearance of the school-room?

8. What can you say of the teacher's responsibility for the health of the child?

9. Would a knowledge of *Mental Philosophy* be a valuable acquisition for a teacher? Why?

10. Should a teacher be held responsible for the intellectual growth of his pupil?

11. Describe the ORDER OF STUDIES to be pursued by the pupil.

12. Should the teacher be responsible for the moral and religious training of his pupil?

13. How should the teacher look upon skepticism? Sectarianism and indifference to moral subjects?

14. What can you say of the personal habits of the teacher?

15. If it be true that "Happy is the man whose habits are his friends," should the teacher look well to his habits and those of his pupils?

16. What should be his order? Courtesy? Punctuality?

17. What should be the *general character* of the teacher?

18. What influence does a knowledge of Trigonometry, Mental and Moral Philosophy, Logic, Rhetoric, Music and "the Languages," have on the mind of the teacher in preparing him for his especial work?

19. What should pupils be taught except what is found in the text-books?

20. What is your view of the modes of teaching?

21. Which is preferable, the *"pouring in,"* or "drawing out process?"

22. Is there a *more excellent way* than either?

23. What would you do to *arouse* and *develope* the *energies* of the *mind?*

24. What can you say in regard to the manner of conducting recitations?

25. Give what you regard as the natural order of presenting any subject.

26. What is the best way of exciting an interest in study?

27. What *incentives* to study would you place before your pupils?

28. Give your idea of *Emulation* in the school.

29. What is your view of offering prizes to scholars?

30. Do prizes usually reward *effort* or *success, worth* or talent?

31. State what you regard as the proper incentives to study.

32. What can you say of *school government?*

33. Mention the elements of a good *disciplinarian.*

34. What *means* would you use to secure *good order?*

35. Is it *wise* or *otherwise* for a teacher to ask a mischievous or vicious pupil to do a favor for him?

36. Would you write out a long set of *rules* by which you propose to govern the school?

37. What can you say of punishment in the school-room?

38. Describe proper and improper punishments.

39. Give your *arrangements* or *programme* for a given school.

40. What can you say in regard to recesses? Also in assigning lessons and reviews?

41. Give in full your opinion of public examinations.

42. What can you say of the relation of the *teacher* to the parents of his pupils?

43. What can you say of the teacher's relation to his profession?

44. What is your view of self-culture and mutual aid of teachers?

45. Give your views of Teachers' Institutes and Associations?

46. What can you say of Physical Education? Mention briefly what different educational forces and processes are required and brought into exercise during the *Objective*, Transition and Subjective periods of the pupils.

47. Give your opinion of recreation.

48. What is implied in *human culture?*

49. What can you say in regard to the education of the *Affections?* The *Will?* *Conscience?*

50. State briefly your idea of the *Science* of *Education.*

CHAPTER XX.
MISCELLANEOUS QUESTIONS.

The following *Test Questions* have been kindly furnished the *author* by several *eminent educators* from widely different parts of the country; questions which *they* themselves *had used* in the *Examination* of *Teachers.* See suggestion at the head of "General Questions," in Chapter XI.

1. *When* and by *whom* was the Continent of South America discovered? And when and by whom was the Continent of North America discovered?

2. *When* and *where* did the Continental Congress meet?

3. Who succeeded Gen. Howe as Commander-in-Chief of the British Army?

4. Name the original thirteen States of the Union.

5. What is a Tariff?

6. What House or Family of Sovereigns occupied the English throne when America was *discovered?* When it was *first settled?* And at the time of the *Declaration of Independence?*

7. What can you say of Mohammed?

8. What was New York first called? Give some account of its first settlement.

9. Who was Anne Hutchinson?

10. Compare the characters and habits of the early Virginia colonists with those of the early New England colonists.

11. Describe the formation and establishment of the Federal Constitution.

12. Describe the naval engagement by Commodore Perry.

13. Describe the Indian War in Florida.

14. Describe the legislative, executive and judicial branches of our National Government.

15. What can you say of England under the Romans?

16. What can you say of Oliver Cromwell?

17. What agency promoted intercourse and civilization in the early periods of history?

18. Whence have the religious institutions and culture of later nations been derived?

19. Under what ruler did the Athenians attain the highest refinement at home and the greatest power abroad?

20. What period of Grecian history is noted for rapid advancement in the practical sciences and philosophy, and who were the most noted men of the time?

21. What seems to have been the cause of the degeneracy and final failure of the Roman Republic?

22. What is the early history of the Anglo Saxons in Britain, and why were they so called?

23. What had the greatest influence upon the development of the Christian Middle Age?

24. What was the origin of the Crusades, and what was their social and religious effect?

25. Who granted the Magna Charta of England, and of what may it be termed the foundation?

26. What great inventions were made in the 14th and 15th Centuries?

27. Is the Constitution a league? If not, explain its nature, and state the reasons.

28. On what authority was it established? And if its existence is subject to any conditions, where are they defined?

29. What was the political status of the English colonies in North America before the Revolution? When and how did they become States?

30. What distribution, and into how many parts does the Constitution make of the powers of this Government?

31. What are the rules by which we are to ascertain the true intent and meaning of any provision in the Constitution?

32. Who is the final Judge or interpreter in controversies concerning it?

33. Does the Constitution provide any means for controlling the vote of a Senator or Representative?

34. State the term of office and mode of electing President, Senator, Representative, and Judge of the Supreme Court, and reasons for the difference.

35. What powers are vested in the United States which do not belong to the States, and, conversely, what exclusive powers have the States?

36. In what manner does the Constitution provide for making war and peace? Mention any instance in which this power may have been usurped.

37. Where is the serous tissue found, and what is its function?

38. Which member of the body affords the greatest evidence of man's superiority over other animals?

39. Describe briefly the structure of the bones.

40. What is the effect upon the bones of severe labor in youth?

41. How are the muscles formed, and how are they attached?

42. In the economy of the muscular system state a striking fact of man's dependence upon a higher power.

43. Name the organs which constitute the digestive apparatus.

44. When labor is lessened what is the effect of continuing the same amount of food as when labor is greater?

45. How is the body sustained when food can not be taken?

46. Is it beneficial to use food which is the most easily digested at all times? Give some reason.

47. How are the arteries and the veins connected?

48. Name the organs used in respiration.

49. Name the organs used in the process of digestion.

50. Describe the circulation of the blood.

51. Trace the food taken into the mouth, through its various changes, until it becomes pure blood.

52. How many different bones in the human system?

· 53. How many pairs of muscles?

54. What is the difference between veins and arteries?

55. What is the difference between venous and arterial blood?

56. Describe the eye.

57. What is the cerebellum and where is it situated?

58. Explain the situation and uses of the diaphragm.

59. Describe a muscle.

60. What is the difference between a nerve of sensation and a nerve of motion?

61. How may a person be deprived of all power of motion, and yet have sensation of feeling?

62. What effect has respiration upon the air breathed?

63. Give the most common abuse by which persons lose health.

64. What is the difference between warm and cold blooded animals?

65. Explain the structure and uses of the spinal column.

66. In cases of injury to blood vessels, how can you tell whether the injury be in an artery or vein?

67. From which does greater danger arise, from the severing of an artery or of a vein? And why?

68. What special provision is made by our Creator for the protection of the arteries?

69. How can you stop the flowing of blood from a severed artery?

70. What is a suture?

71. What difference between the skull of an infant and of an adult?

72. What conditions as to warmth and ventilation, are most conducive to health?

73. What are the constituent elements of Atmospheric air?

74. What is Specific Gravity?

75. What is the effect of respiration on the blood?

76. Explain the principle of the Thermometer, Barometer and Air Pump. .

77. What is a body?

78. What are the essential properties of matter?

79. Define and illustrate impenetrability?

80. Define Gravity. What is Inertia?

81. Illustrate by examples the different laws of Inertia.

82. What is the fundamental law of Mechanics?

83. Name the Mechanical Powers.

84. Give a formula for ascertaining the power of the lever.

85. What is Pneumatics? State the law of falling bodies.

86. In what respect do liquids differ from solids?

87. How do you ascertain the weight of the Atmosphere?

88. What is Specific Gravity? How do you ascertain the Specific Gravity of solids, and upon what principles is your rule based?

89. The flash of a gun is seen by you $3\frac{1}{2}$ seconds before the report is heard. At what distance is the gun from you?

90. What circumstances are favorable to safety during a Thunder Storm?

91. Why can you not see an object through a tube bent at right angles?

92. Describe the eye.

93. Name the different kinds of lenses and the effect of each upon parallel rays of light passing through it.

94. Why does a body of water appear less deep than it really is?

95. What is Galvanism? Of what does Mechanics treat?

96. How does the Barometer help us to determine the elevation of a mountain?

97. What is the rule for finding the specific gravity of bodies lighter than water?

98. What is Specific Gravity? Describe the Pulley and its uses.

99. What is the Leyden Jar? Hydrostatic Paradox?

100. What is the greatest height of a column of water sustained by atmospheric pressure only; and what would be the result from a change of temperature?

101. Why does salt cause ice to become fluid?

102. If a body be thrown directly upward, and return again to the earth in eleven seconds, what height does the body reach?

103. Why do the particles of Water when left to themselves, have a tendency to assume a globular form, as is noticed in rain drops and tears?

104. Rule for finding Specific Gravity.

105. Define Language in its most extended sense, and explain the office of English Grammar in reference to it.

106. Name and define the elements of the following sentence:

"Ah! the good boy and his dog run rapidly down the hill."

107. Parse the words italicised in the following sentence:

"I do not know *what is to be done*."

108. Correct in every respect the following sentences, (if correction be needed,) giving the reason, founded upon a grammatical principle:

a. "The Book laid on the floor."
b. "I done the sum on page forty six."
c. "The stream has overflown its banks."
d. "There is no study in our schools so hard to teach as Grammar."

109. Write the possessive plural of *lady* and *man*.

110. Define abstract nouns, and give five examples.

111. Compare the adjectives *evil, little, front, much, brilliant.*

112. Decline the pronoun *which.*

113. Give the principal parts of the verbs *strike, forget, pay, hear.*

114. What is the difference between the *voice* of a verb and its *mode*?

115. Write the conjugation of the verb *be*, indicative mode, past tense.

116. Write a sentence in which the grammatical predicate is modified by a verb in the infinitive.

117. Parse the italicised words in the following sentences:

"He gave *me what* I *desire*." "Milton, the *poet*, was *blind*."

118. Analyze the following lines:

"Now came still evening on, and twilight gray
Had in her sober livery all things clad."

119. Of what does Mathematical Geography treat? Physical Geography?

120. Political Geography? Name the three largest States in the Union, and the three smallest, with the capital of each.

121. A and B travelled around the earth,—A on the parallel of Chicago, and B on the parallel of London. Which travelled the greatest number of miles? How many degrees did each travel?

122. Give the boundaries of Ireland.

123. What is the length of the longest day in latitude 74 degrees?

124. What is the width of the Torrid Zone, in English miles?

125. Give the situation of Havre, Montevideo, Samarcand, Odessa, Caraccas.

126. Name the countries in Europe that border on the Mediterranean.

127. Describe a water voyage from Odessa to St. Petersburg, naming all the waters through which you would pass.

128. Draw an outline map of Asia.

129. Sketch a map of Wisconsin, with the principal Towns, their Railroad connections, Lakes and Rivers.

130. Bound your own township.

131. How does Wisconsin compare in size with France?

132. Why are the tropics $23\frac{1}{2}$ degrees from the equator?

133. Upon what cau. es does the temperature of a country depend?

134. What causes the saltness of the Ocean?

135. From what sources do Springs and Water Veins receive their support?

136. What are the prevailing Winds of the Mississippi Valley, and why?

137. What circumstances chiefly affect the climate of a country?

138. Bound Texas.

139. What is meant by "Water Shed," as used in Geography?

140. Name the principal Ports on the Pacific coast of North America.

141. Of what States does the river Mississippi form a part of the boundary?

142. How is Italy bounded? Name its principal mountains, rivers, and political division.

143. Explain the process for the Division of vulgar fractions, and the reason of the common rule therefor.

144. Find the interest on $355.56 for 4 yrs. 8 mo. 24d., at 5 per cent.

145. Divide $1000 among A, B and C, so that A will receive $120 more than B, and B $95 more than C.

146. How much cotton at $2\frac{1}{2}$ cents per lb. can be bought for $2,500, deducting brokerage at the rate of $2\frac{1}{2}$ per cent. on the amount purchased?

147. If 4 men dig a trench 20 ft. long, $12\frac{1}{2}$ ft. deep, and 2 ft. wide, in 4 days, working 10 hours a day, how many men will dig a trench 10 ft. long, $6\frac{1}{4}$ ft. deep, and $3\frac{1}{2}$ ft. wide, in 7 days, working 3 hours per day?

148. What is the square root of 4.8681?

149. When and how do ciphers give value to significant figures, as integers or decimals? What is Subtraction?

150. What is the difference between Reduction of Denominate Numbers and Reduction of Fractions?

151. When you add Fractions, why not add Denominators as well as Numerators?

152. Perform the following operation: $.04 \div .005 \times .3$, and give the reason for pointing off.

153. If you buy a slate for 8 cents, and sell it for 10 cents, what per cent. do you make?

154. What is the interest on $24,000 for $\frac{3}{4}$ of an hour, at 10 per cent. per annum?

155. Why and how are fractions reduced to a common denominator?

156. Find the entire quotient arising from dividing two thousand and one millionths by one hundredth.

157. If I have a piece of land 16$\frac{2}{3}$ rods long, and 3$\frac{1}{2}$ rods wide, what will be the length of another piece that is 7 rods long, and contains the same area?

158. If 82 men build a wall 36 ft. long, 8 ft. high, and 4 ft. thick, in 4 days, in what time will 48 men build a wall 864 ft. long, 6 ft. high, and 3 ft. thick?

159. In a compound partnership how would you find each partner's share of the gain or loss?

160. A capitalist has $25,000; he invests 20 per cent. in bank stock, 37$\frac{1}{2}$ per cent. in railroad stock, and the remainder in government bonds; what per cent. and what sum did he invest in the bonds?

161.   CHICAGO, Jan. 1st, 1860.
$382.50.

For value received I promise to pay, on the 10th day of June next, to S. Brooks or order, the sum of three hundred and eighty-two dollars and fifty cents, with interest from date, at 7 per cent.

                                        J. DAVIS.

Required the amount of the above note at the time of settlement.

162. Bought 24 bbls. of flour for $168, and sold $\frac{2}{3}$ of it at $6.75 per bbl., and the remainder at $7.50 per bbl. Did I gain or lose, and how much?

163. If gold 18, 21, 17, 19, 20 carats fine be melted together, what will be the fineness of the compound?

164. In the center of a square garden there is an artificial circular pond, covering an area of 810 sq. ft., which is one-tenth of

the whole garden; how many rods of fence will enclose the garden?

165. What is the difference between Algebra and Arithmetic?

166. What is an equation?

167. If $-a \times -a = +a^2$, explain the reason for the change of signs.

168. What is Algebra?

169. Write the symbols generally used in Algebraic formulas, and give the signification of each.

170. Name the axioms. What is the meaning of Transposition, and what is the occasion of its use?

171. Find the value of x in the equation $\frac{x}{4} + \frac{2x}{5} - 3 = \frac{x}{2} + 6$.

172. Solve $\frac{3x}{4} + \frac{x-5}{3} = \frac{x}{2} - \frac{x-5}{3} + 13$.

173. What is Elimination? How many equations are requisite to the solution of a question involving two or more unknown quantities?

174. Solve by elimination by addition or subtraction,
$$x + 5y = 16.$$
$$3x - 2y = -3.$$

175. Solve by Substitution, $3x - 4y = 8.$
$$2x - 3y = 5.$$

176. Solve by Comparison, $3\frac{1}{2}x + 4\frac{3}{4}y = 21.$
$$5x + 2y = 16.$$

177. What is a complete quadratic equation? What is an incomplete quadratic?

178. Solve the following: $x^2 \div 5x = 36.$

179. Solve the following: $3x^2 + {}^2x + 8 = 41.$

180. State and solve the following: A and B find $120. Had A found $10 more, and B $10 less, they would have found equal sums. What did each find?

181. State and solve the following: Three numbers are equal to 280; one-third of the second equals the first, and the third is 50 less than one-half the second. What is each number?

182. If a certain number be increased by one-half itself and

the sum be multiplied by the number, the product will be equal to 7 times the number less 4. What is the number?

$$\frac{x}{2} + \frac{27}{3} + z = 4\frac{5}{8}.$$

$$x + y + 3z = 4\frac{5}{8}.$$

183. Prove that *minus* multiplied or divided by *minus* gives *plus*.

184. Find the greatest common divisor of $x^4 + a^2x^2 + a^4$ and $x^4 + ax^3 - a^2x - a^4$.

185. If the difference of two fractions is equal to $\frac{p}{q}$ show that p times their sum is equal to q times the difference of their squares.

186. Find the value of x in $\frac{x}{a} + 5x - 1 = c$.

187. Find two numbers such that their sum, their product, and the difference squares shall be all equal to each other.

188. What is the present value of an annuity of $112.50, to commence at the end of 10 years, and to continue 20 years at 4 per cent.?

189. What are the different modes of computing logarithms? Which is the most convenient, and why?

190. Show that the square described on the difference of two lines is equivalent to the sum of the squares described on the lines diminished by twice the rectangle contained by the lines. Give the Algebraic expression of this proposition.

191. Show that like powers or roots of proportional magnitudes are proportional.

192. Show that the distance between any two points on the surface of a sphere is less if measured on the arc of a great circle than on the arc of a small circle.

193. Show that the sum of the three sides of a spherical triangle is less than the circumference of a great circle.

194. Make the diagram of any arc greater than 90 degrees, with its sine, cosine, &c., designating each of the parts below the diagram with letters.

195. Give the modes for finding any unknown parts of a right-angled triangle.

196. Find the formula for the sine of the difference of two angles or arc..

197. Give *the three equations* which are the *primary formulas* of *spherical trigonometry.*

198. How many modes of measuring lands? Describe each process.

199. Give a brief account of the different systems of Astronomy and Kepler's laws.

200. Explain the variation of the seasons and of the length of days and nights.

201. Give the process of solving the sun's parallax and what is known as to the parallax of any fixed star.

202. How may we calculate the quantity of a solar eclipse at a particular place?

203. What is the use of **Axioms** in Geometry?

204. Describe four different **Quadrilaterals.** Three different kinds of Triangles.

205. What is **Plane Geometry?** Solid Geometry? A Straight Line?

206. How many right angles can be formed about a given point?

207. Demonstrate, and use **Algebraic symbols,** if desired:

The angles of a triangle are together equal to two right angles.

208. When two straight lines cross each other the opposite or vertical angles are equal.

209. If a straight line meet two parallel straight lines, the sum of the interior angles on the same side will be equal to two right angles.

210. If all the sides of a polygon are produced in the same direction, the sum of the exterior angles will be equal to four right angles.

211. Two diagonals of any parallelogram mutually bisect each other.

212. Any two triangles having two angles and the included side of the one equal to two angles and the included side of the other, each to each, are equal.

213. A triangle is equivalent to half a parallelogram of equal base and altitude.

214. If a straight line be divided into two parts, the square described upon the whole line is equal to the sum of the squares described upon each of the parts, increased by twice the rectangle formed by the two parts. Formula :—$(x+y)^2 = x^2 + y^2 + 2xy.$

215. An inscribed angle is measured by half the arc included between its sides.

216. The angles inscribed in the same segment of a circle are equal.

217. The radius perpendicular to a chord bisects the chord and its subtended arc.

218. To find the center of a given circle.

219. Parallelograms of equal altitude are to each other as their bases.

220. In equal circles, equal arcs are subtended by equal chords.

221. If we wish to build a hexagonal brick house, at what angle shall we make the corner brick? Give the reason for the conclusion.

222. How many circumstances, and what are they, under which triangles are alike?

223. How will you occupy the attention and employ the time of the younger class of pupils while they are in the school-room?

224. How will you teach the Alphabet?

225. Will you allow pupils to "spell out" words in reading, and to what extent?

226. What will you recommend for the first study of a child who has learned to read well enough to take up a regular study?

227. Enumerate some of the things that you will teach orally.

228. Give a schedule of the course of study you will recommend from the time the pupil commences the study of books to his fourteenth year.

229. What special means will you employ to secure the interest and co-operation of parents?

230. At what degree of temperature will you keep your school-room?

231. What incentives will you use in the school-room?

232. What plan have you for controlling whispering?

233. What plan have you for securing punctuality?

234. What should be the aim of all human culture?

FINIS.

ORTHOGRAPHY AND READING.

NATIONAL SERIES

OF

READERS AND SPELLERS,

BY PARKER & WATSON.

The National Primer · · · · · · · · · · .$ 25

National First Reader · · · · · · · · . 38

National Second Reader · · · · · · · · 63

National Third Reader · · · · · · · · · 95

National Fourth Reader · · · · · · · · .1 50

National Fifth Reader · · · · · · · · · .1 88

National Elementary Speller · · · · · · · 25

National Pronouncing Speller · · · · · · 45

Independent Third Reader · · · · · · · ·

Independent Fourth Reader · · · · · · · 95

Independent Fifth Reader · · · · · · · .1 50

The salient features of these works which have combined to render them so popular may be briefly recapitulated as follows :

1. THE WORD BUILDING SYSTEM.—This famous progressive method for young children originated and was copyrighted with these books. It constitutes a process by which the beginner with *words* of one letter is gradually introduced to additional lists formed by prefixing or affixing single letters, and is thus led almost insensibly to the mastery of the more difficult constructions. This is justly regarded as one of the most striking modern improvements in methods of teaching.

2. TREATMENT OF PRONUNCIATION.—The wants of the youngest scholars in this department are not overlooked. It may be said that from the first lesson the student by this method need never be at a loss for a prompt and accurate rendering of every word encountered.

3. ARTICULATION AND ORTHOEPY are recognized as of primary portance.

ORTHOGRAPHY AND READING—Continued.

4. PUNCTUATION is inculcated by a series of interesting *reading lessons.* the simple perusal of which suffices to fix its principles indelibly upon the mind.

5. ELOCUTION. Each of the higher Readers (3d, 4th and 5th) contains elaborate, scholarly, and thoroughly practical treatises on elocution. This feature alone has secured for the series many of its warmest friends.

6. THE SELECTIONS are the crowning glory of the series. Without exception it may be said that no volumes of the same size and character contain a collection so diversified, judicious, and artistic as this. It embraces the choicest gems of English literature, so arranged as to afford the reader ample exercise in every department of style. So acceptable has the taste of the authors in this department proved, not only to the educational public but to the reading community at large, that thousands of copies of the Fourth and Fifth Readers have found their way into public and private libraries throughout the country, where they are in constant use as manuals of literature, for reference as well as perusal.

7. ARRANGEMENT. The exercises are so arranged as to present constantly alternating practice in the different styles of composition, while observing a definite plan of progression or gradation throughout the whole. In the higher books the articles are placed in formal sections and classified topically, thus concentrating the interest and inculcating a principle of association likely to prove valuable in subsequent general reading.

8. NOTES AND BIOGRAPHICAL SKETCHES. These are full and adequate to every want. The biographical sketches present in pleasing style the history of every author laid under contribution.

9. ILLUSTRATIONS. These are plentiful, almost profuse, and of the highest character of art. They are found in every volume of the series as far as and including the Third Reader.

10. THE GRADATION is perfect. Each volume overlaps its companion preceding or following in the series, so that the scholar, in passing from one to another, is barely conscious, save by the presence of the new book, of the transition.

11. THE PRICE is reasonable. The books were not *trimmed* to the minimum of size in order that the publishers might be able to denominate them " the cheapest in the market," but were made *large enough* to cover and suffice for the grade indicated by the respective numbers. Thus the child is not compelled to go over his First Reader twice, or be driven into the Second before he is prepared for it. The competent teachers who compiled the series made each volume just what it should be, leaving it for their brethren who should use the books to decide what constitutes true *cheapness.* A glance'over the books will satisfy any one that the same amount of matter is nowhere furnished at a price more reasonable. Besides which another consideration enters into the question of relative economy, namely, the

12. BINDING. By the use of a material and process known only to themselves, in common with all the publications of this house, the National Readers are warranted to out-last any with which they may be compared—the ratio of relative durability being in their favor as two to one.

Parker & Watson's National Series of Readers.

TESTIMONIALS.

From Hon. T. A. Parker, *State Sup't of Public Instruction, Missouri.*

By authority of law it becomes my duty to recommend a list of Text-books for use in the Public Schools of Missouri. I deem it necessary to approve a list of books which will secure to the youth of the State a *uniform, cheap, and practical* course of study, and after careful examination have selected the following: The National Readers and Spellers, *Monteith & McNally's Geographies, Peck's Ganot's Natural Philosophy, Jarvis' Physiology and Health,* &c., &c.

From Sam'l P. Bates, LL.D., *Asst. Supt. Public Schools of Pennsylvania.*

I find that your series of Parker & Watson's National Readers are going into use in all our leading Normal Schools. They are unquestionably ahead of any thing yet published.

From A. J. Haile, *Prin. Hebrew Educational Institute, Memphis, Tennessee.*

I take great pleasure in bearing testimony to the superior merits of Parker & Watson's Series of " National Readers."

From Prof. F. S. Jewell, *of the New York State Normal School.*

It gives me pleasure to find in the National Series of School-Readers ample room for commendation From a brief examination, I am led to believe that we have none equal to them. I hope they will prove as popular as they are excellent.

From Moses T. Brown, *Superintendent Public Schools, Toledo, Ohio.*

The different Series of other authors were critically examined by our Board of Education and myself, and the decision was unanimous in favor of the National Series. Our teachers are delighted with the books, and none more so than our primary teachers. *I consider the Series better adapted to our graded school system than any other now before the public.*

From Wm. B. Ames, *Superintendent of Schools, Morris, Connecticut.*

They are well adapted to all degrees of scholarship—one lesson prepares the mind of the pupil for the next in consecutive order, from book to book—till the highest order of English composition is attained in the Fifth Reader.

From John S. Hart, *Prin. N. J. State Normal School.*

I approve of Parker & Watson's Readers highly. The selections are judicious, the arrangement good, and the books well made mechanically. We have adopted the 3d, 4th, and 5th of the Series in this school.

From R. P. Deckard, *President Ewing College, La Grange, Texas.*

I think the National Series of Readers the best I have seen.

Extracts from Report made to the California State Teachers' Association.

The Committee, in presenting to this Convention the Series of Readers by Parker & Watson, would state that, regarded as a whole, we would give our unqualified support to them in preference to all others.

From B. J. Young, *Superintendent Schools, Shelbyville, Illinois.*

The National Readers have been selected for use in the public schools of this city, and are giving very excellent satisfaction. During ten years' experience in teaching, I have found no books so well adapted to secure rapid and thorough progress.

From the Wilmington (N. C.) Daily Herald.

The National Series has attained probably a higher reputation than any other complete series of School-Books in existence.

☞ For further testimony of a similar character, see special circular, or current numbers of the Educational Bulletin

The National Readers and Spellers.

THEIR RECORD.

These books have been adopted by the School Boards, or official authority, of the following important States, cities, and towns—in most cases for exclusive use.

The State of Minnesota.
The State of Missouri.

New York.
New York City.
Brooklyn.
Buffalo.
Albany.
Rochester.
Troy.
Syracuse.
Elmira.
&c., &c.

Pennsylvania.
Reading.
Lancaster.
Erie.
Scranton.
Carlisle.
Carbondale.
Meadville.
Schuylkill Haven.
Williamsport.
Norristown.
Bellefonte.
Altoona.
&c., &c.

New Jersey.
Newark.
Jersey City.
Paterson.
Trenton.
Camden.
Elizabeth.
New Brunswick.
Phillipsburg.
Orange.
&c., &c.

Delaware.
Wilmington.

D. C.
Washington.

Illinois.
Chicago.
Peoria.
Alton.
Springfield.
Aurora.
Galesburg.
Rockford.
Rock Island.
&c., &c.

Wisconsin.
Milwaukee.
Fond du Lac.
Oshkosh.
Janesville.
Racine.
Watertown.
Sheboygan.
La Crosse.
Waukesha.
Kenosha.
&c., &c.

Michigan.
Grand Rapids.
Kalamazoo.
Adrian.
Jackson.
Monroe.
Lansing.
&c., &c.

Ohio.
Toledo.
Sandusky.
Conneaut.
Chardon.
Hudson.
Canton.
Salem.
&c., &c.

Indiana.
New Albany.
Fort Wayne.
Lafayette.
Madison.
Logansport.
&c., &c.

Iowa.
Davenport.
Burlington.
Muscatine.
Mount Pleasant.
&c.

California.
Sacramento.
Marysville.
&c.

Oregon.
Portland.
Salem.
&c.

* Virginia.
Richmond.
Norfolk.
Petersburg.
Lynchburg.
&c.

* Carolina.
Wilmington.
Charleston.

* Georgia.
Savannah.

* Louisiana.
New Orleans.

* Tennessee,
Memphis.

* With the organization and progress of common-school systems at the South, this list will, of course, be greatly increased. These points are, at present, almost the only ones enjoying the advantages of public schools. The National Readers are used in large number of the best private schools and academies throughout the South.

The *Educational Bulletin* records periodically all new points gained.

SCHOOL-ROOM CARDS,

To Accompany the National Readers.

Eureka Alphabet Tablet *1 50

Presents the alphabet upon the Word Method System, by which the
child will learn the alphabet in nine days, and make no small progress in
reading and spelling in the same time.

National School Tablets, 10 Nos. *7 50

Embrace reading and conversational exercises, object and moral lessons, form, color, &c. A complete set of these large and elegantly illustrated Cards will embellish the school-room more than any other article
of furniture.

READING.

Fowle's Bible Reader $1 00

The narrative portions of the Bible, chronologically and topically arranged, judiciously combined with selections from the Psalms, Proverbs,
and other portions which inculcate important moral lessons or the great
truths of Christianity. The embarrassment and difficulty of reading the
Bible itself, by course, as a class exercise, are obviated, and its use made
feasible, by this means.

North Carolina First Reader 40
North Carolina Second Reader 65
North Carolina Third Reader 1 00

Prepared expressly for the schools of this State, by C. H. Wiley, Superintendent of Common Schools, and F. M. Hubbard, Professor of Literature in the State University.

Parker's Rhetorical Reader 1 00

Designed to familiarize Readers with the pauses and other marks in
general use, and lead them to the practice of modulation and inflection of
the voice.

Introductory Lessons in Reading and Elo-
cution 75

Of similar character to the foregoing, for less advanced classes.

High School Literature 1 50

Admirable selections from a long list of the world's best writers, for exercise in reading, oratory, and composition. Speeches, dialogues, and
model letters represent the latter department.

ORTHOGRAPHY.

SMITH'S SERIES

Supplies a speller for every class in graded schools, and comprises the most complete and excellent treatise on English Orthography and its companion branches extant.

1. Smith's Little Speller · . . . . $ 20

First Round in the Ladder of Learning.

2. Smith's Juvenile Definer , . . . 45

Lessons composed of familiar words grouped with reference to similar signification or use, and correctly spelled, accented, and defined.

3. Smith's Grammar-School Speller · . . 50

Familiar words, grouped with reference to the sameness of sound of syllables differently spelled. Also definitions, complete rules for spelling and formation of derivatives, and exercises in false orthography.

4. Smith's Speller and Definer's Manual . 90

A complete *School Dictionary* containing 14,000 words, with various other useful matter in the way of Rules and Exercises.

5. Smith's Hand-Book of Etymology . . 1 25

The first and only Etymology to recognize the *Anglo-Saxon* our *mother tongue;* containing also full lists of derivatives from the Latin, Greek, Gaelic, Swedish, Norman, &c., &c.; being, in fact, a complete etymology of the language for schools.

Sherwood's Writing Speller · 15

Sherwood's Speller and Definer · . . . 15

Sherwood's Speller and Pronouncer . . . 15

The Writing Speller consists of properly ruled and numbered blanks to receive the words dictated by the teacher, with space for remarks and corrections. The other volumes may be used for the dictation or ordinary class exercises.

Price's English Speller *15

A complete spelling-book for all grades, containing more matter than "Webster," manufactured in superior style, and sold at a lower price—consequently the cheapest speller extant.

Northend's Dictation Exercises 63

Embracing valuable information on a thousand topics, communicated in such a manner as at once to relieve the exercise of spelling of its usual tedium, and combine it with instruction of a general character calculated to profit and amuse.

Wright's Analytical Orthography 25

This standard work is popular, because it teaches the elementary sounds in a plain and philosophical manner, and presents orthography and orthoepy in an easy, uniform system of analysis or parsing.

Fowle's False Orthography 45

Exercises for correction.

Page's Normal Chart *3 75

The elementary sounds of the language for the school-room walls.

ENGLISH GRAMMAR.

CLARK'S DIAGRAM SYSTEM.

Clark's First Lessons in Grammar . . . 45
Clark's English Grammar 1 00
Clark's Key to English Grammar 75
Clark's Analysis of the English Language . 60
Clark's Grammatical Chart *3 75

The theory and practice of teaching grammar in American schools is meeting with a thorough revolution from the use of this system. While the old methods offer proficiency to the pupil only after much weary plodding and dull memorizing, this affords from the inception the advantage of *practical Object Teaching*, addressing the eye by means of illustrative figures; furnishes association to the memory, its most powerful aid, and diverts the pupil by taxing his ingenuity. Teachers who are using Clark's Grammar uniformly testify that they and their pupils find it the most interesting study of the school course.

Like all great and radical improvements, the system naturally met at first with much unreasonable opposition. It has not only outlived the greater part of this opposition, but finds many of its warmest admirers among those who could not at first tolerate so radical an innovation. All it wants is an impartial trial, to convince the most skeptical of its merit. No one who has fairly and intelligently tested it in the school-room has ever been known to go back to the old method. A great success is already established, and it is easy to prophecy that the day is not far distant when it will be the *only system of teaching English Grammar.* As the SYSTEM is copyrighted, no other text-books can appropriate this obvious and great improvement.

Welch's Analysis of the English Sentence . 1 00

Remarkable for its new and simple classification, its method of treating connectives, its explanations of the idioms and constructive laws of the language, &c.

ETYMOLOGY.

Smith's Complete Etymology, ... 1 25

Containing the Anglo-Saxon, French, Dutch, German, Welsh, Danish, Gothic, Swedish, Gaelic, Italian, Latin, and Greek Roots, and the English words derived therefrom accurately spelled, accented, and defined.

The Topical Lexicon, ... 1 75

This work is a School Dictionary, an Etymology, a compilation of synonyms, and a manual of general information. It differs from the ordinary lexicon in being arranged by topics instead of the letters of the alphabet, thus realizing the apparent paradox of a "Readable Dictionary." An unusually valuable school-book.

9

Clark's Diagram English Grammar.

TESTIMONIALS.

From J. A. T. DURNIN, Principal Dubuque R. C. Academy, Iowa.

In my opinion, it is well calculated by its system of analysis to develop those rational faculties which in the old systems were rather left to develop themselves, while the memory was overtaxed, and the pupils discouraged.

From B. A. COX, School Commissioner, Warren County, Illinois.

I have examined 150 teachers in the last year, and those having studied or taught Clark's System have universally stood fifty per cent. better examinations than those having studied other authors.

From M. H. B. BURKET, Principal Masonic Institute, Georgetown, Tennessee.

I traveled two years amusing myself in instructing (exclusively) Grammar classes with Clark's system. The first class I instructed fifty days, but found that this was more time than was required to impart a theoretical knowledge of the science. During the two years thereafter I instructed classes only *thirty* days each. Invariably I proposed that unless I prepared my classes for a more thorough, minute, and accurate knowledge of English Grammar than that obtained from the ordinary books and in the ordinary way in from one to two years, I would make no charge. I never failed in a solitary case to far exceed the hopes of my classes, and made money and character rapidly as an instructor.

From A. B. DOUGLASS, School Commissioner, Delaware County, New York.

I have never known a class pursue the study of it under a *live* teacher, that has not succeeded; I have never known it to have an opponent in an educated teacher who had *thoroughly* investigated it; I have never known an *ignorant* teacher to examine it; I have never known a teacher who has used it, to try any other.

From J. A. DODGE, Teacher and Lecturer on English Grammar, Kentucky.

We are tempted to assert that it foretells the dawn of a brighter age to our mother tongue. Both pupil and teacher can fare sumptuously upon its contents, however highly they may have prized the manuals into which they may have been initiated, and by which their expressions have been moulded.

From W. T. CHAPMAN, Superintendent Public Schools, Wellington, Ohio.

I regard Clark's System of Grammar the best published. For teaching the analysis of the English Language, it surpasses any I ever used.

From F. S. LYON, Principal South Norwalk Union School, Connecticut.

During ten years' experience in teaching, I have used six different authors on the subject of English Grammar. I am fully convinced that Clark's Grammar is better calculated to make thorough grammarians than any other that I have seen.

From CATALOGUE OF ROHRER'S COMMERCIAL COLLEGE, St. Louis, Missouri.

We do not hesitate to assert, without fear of successful contradiction, that a better knowledge of the English language can be obtained by this system in six weeks than by the old methods in as many months.

From A. PICKETT, President of the State Teachers' Association, Wisconsin.

A thorough experiment in the use of many approved authors upon the subject of English Grammar has convinced me of the superiority of Clark. When the pupil has completed the course, he is left upon a foundation of *principle*, and not upon the *dictum* of the author.

From GEO. P. McFARLAND, Prin. McAllisterville Academy, Juniata Co., Penn.

At the first examination of public-school teachers by the county superintendent, when one of our student teachers commenced analyzing a sentence according to Clark, the superintendent listened in mute astonishment until he had finished, then asked what that meant, and finally, with a very knowing look, said such work wouldn't do here, and asked the applicant to parse the sentence right, and gave the lowest certificates to all who barely mentioned Clark. Afterwards, I presented him with a copy, and the next fall he permitted it to be partially used, while the third or last fall, he openly commended the system, and appointed three of my best teachers to explain it at the two Institutes and one County Convention held since September.

☞ For further testimony of equal force, see the Publishers' Special Circular, or current numbers of the Educational Bulletin.

GEOGRAPHY.

THE

NATIONAL GEOGRAPHICAL SYSTEM.

I. Monteith's First Lessons in Geography, $ 35
II. Monteith's Introduction to the Manual, · 63
III. Monteith's New Manual of Geography, · 1 00
IV. Monteith's Physical & Intermediate Geog. 1 70
V. McNally's System of Geography, · · · 1 88

Monteith's Wall Maps (per set) · · · *20 00

1. PRACTICAL OBJECT TEACHING. The infant scholar is first introduced to *a picture* whence he may derive notions of the shape of the earth, the phenomena of day and night, the distribution of land and water, and the great natural divisions, which mere words would fail entirely to convey to the untutored mind. Other pictures follow on the same plan, and the child's mind is called upon to grasp no idea without the aid of a pictorial illustration. Carried on to the higher books, this system culminates in No. 4, where such matters as climates, ocean currents, the winds, peculiarities of the earth's crust, clouds and rain, are pictorially explained and rendered apparent to the most obtuse. The illustrations used for this purpose belong to the highest grade of art.

2. CLEAR, BEAUTIFUL, AND CORRECT MAPS. In the lower numbers the maps avoid unnecessary detail, while respectively progressive, and affording the pupil new matter for acquisition each time he approaches in the constantly enlarging circle the point of coincidence with previous lessons in the more elementary books. In No. 4, the maps embrace many new and striking features. One of the most effective of these is the new plan for displaying on each map the relative sizes of countries not represented, thus obviating much confusion which has arisen from the necessity of presenting maps in the same atlas drawn on different scales. The maps of No. 5 have long been celebrated for their superior beauty and completeness. This is the only school-book in which the attempt to make a *complete* atlas *also clear and distinct*, has been successful. The map *coloring* throughout the series is also noticeable. Delicate and subdued tints take the place of the startling glare of inharmonious colors which too frequently in such treatises dazzle the eyes, distract the attention, and serve to overwhelm the names of towns and the natural features of the landscape.

11

GEOGRAPHY—Continued.

3. **THE VARIETY OF MAP EXERCISE.** Starting each time from a different basis, the pupil in many instances approaches the same fact no less than *six times*, thus indelibly impressing it upon his memory. At the same time this system is not allowed to become wearisome—the extent of exercise on each subject being graduated by its relative importance or difficulty of acquisition.

4. **THE CHARACTER AND ARRANGEMENT OF THE DESCRIPTIVE TEXT.** The cream of the science has been carefully culled, unimportant matter rejected, elaboration avoided, and a brief and concise manner of presentation cultivated. The orderly consideration of topics has contributed greatly to simplicity. Due attention is paid to the facts in history and astronomy which are inseparably connected with, and important to the proper understanding of geography—and *such only* are admitted on any terms. In a word, the National System teaches geography as a science, pure, simple, and exhaustive.

5. **ALWAYS UP TO THE TIMES.** The authors of these books, editorially speaking, never sleep. No change occurs in the boundaries of countries, or of counties, no new discovery is made, or railroad built, that is not at once noted and recorded, and the next edition of each volume carries to every school-room the new order of things.

6. **SUPERIOR GRADATION.** This is the only series which furnishes an available volume for every possible class in graded schools. It is not contemplated that a pupil must necessarily go through every volume in succession to attain proficiency. On the contrary, *two* will suffice, but *three* are advised ; and if the course will admit, the whole series should be pursued. At all events, the books are at hand for selection, and every teacher, of every grade, can find among them one *exactly suited* to his class. The best combination for those who wish to abridge the course consists of Nos. 1, 3, and 5, or where children are somewhat advanced in other studies when they commence geography, Nos. 2, 3, and 5. Where but *two* books are admissible, Nos. 2 and 4, or Nos. 3 and 5, are recommended.

7. **FORM OF THE VOLUMES AND MECHANICAL EXECUTION.** The maps and text are no longer unnaturally divorced in accordance with the time-honored practice of making text-books on this subject as inconvenient and expensive as possible. On the contrary, all map questions are to be found on the page opposite the map itself, and each book is complete in one volume. The mechanical execution is unrivalled. Paper and printing are everything that could be desired, and the binding is—A. S. Barnes and Company's.

Ripley's Map Drawing $1 25

This system adopts the circle as its basis, abandoning the processes by triangulation, the square, parallels, and meridians, &c., which have been proved not feasible or natural in the development of this science. Success seems to indicate that the circle " has it."

National Outline Maps (per set). $15 00

For the school-room walls. In preparation.

12

Monteith & McNally's National Geographies.

TESTIMONIALS.

From O. F. RUSSELL, *Principal Normal Academy, Arkansas.*

Before the war I used Monteith and McNally's Geographies, and do not hesitate to pronounce them the best that ever came under my observation.

From HAMILTON McAFEE, *Superintendent Caldwell County, Missouri.*

Monteith and McNally's Geographies are superior to any in use; in point of mechanical execution they are certainly unrivaled.

From E. W. PEET, *Principal Walworth County Institute, Wisconsin.*

We have used Monteith's Geographies in the Institute for two years, where they have given the best satisfaction. I most heartily indorse and recommend them.

From JOSEPH PEABODY, *Principal Moody School, Lowell, Massachusetts.*

I have examined Monteith's Geographies carefully, and feel confident in saying that they are *the best series of Geographies* that have come under my observation.

From REV. B. ST. JAMES FRY, A.M., *President Worthington Female College, Ohio.*

We have used McNally and Monteith's Geographies for three years, and would not exchange them for any others in the market.

From R. M. MAGEE, *Superintendent Centre County, Pennsylvania.*

Monteith and McNally's Geographies have been examined carefully, and I am free to say I think they are superior in many respects to any other system.

From W. L. ALEXANDER, *President Nacogdoches College, Texas.*

I regard as perfect, in every respect, "Monteith and McNally's Geographical Series."

From C. P. DE HASS, *Principal North Hill Public School, Burlington, Iowa.*

I favored the adoption of Monteith and McNally's Series of Geographies, because I liked them; and now, after nearly a year's trial of them in the school-room, I like them better than ever.

From R. A. ADAMS, *Member of Board of Education, New York.*

I have found, by examination of the Book of Supply of our Board, that considerably the largest number of any series now used in our public schools is the National, by Monteith and McNally.

From JOSIAH T. READE, *Principal Union School, Marshall, Michigan.*

This series was adopted after a careful examination of the best works in this branch of study, and a year's experience makes us better and better satisfied with our choice.

From EMORY F. STRONG, *Principal High School, Bridgeport, Connecticut.*

We are using, with very great satisfaction, in the school with which I am connected, Monteith and McNally's Geographies. Other schools in this city are using them with the same favorable opinion of their merits.

From A. R. McGILL, *Superintendent Nicollet County, Minnesota.*

I am happy to express my hearty approval of the *Series throughout.*

From JAMES N. TOWNSEND, *Superintendent Public Schools, Hudson, New York.*

I have *carefully* examined your series of Monteith and McNally's Geographies. They are comprehensive, accurate, well graded, handsomely gotten up, *complete.* I am frank to confess that I consider it by far the best series ever published in this country. They are unrivaled. We have recently adopted them in the public school of this city, to the infinite delight of the students, and to the entire satisfaction of the teachers.

☞ For further testimony of similar character, see the Publishers' special circular, or current numbers of the Educational Bulletin.

The National System of Geography,

By Monteith & McNally.

These popular text-books have been adopted, by official authority, for the schools of the following States, cities, and associations—in most cases for exclusive and uniform use.

MINNESOTA. By State Board of Education.
VERMONT. Do. do.
ALABAMA. Do. do.
MISSOURI. By State Superintendent of Common Schools.
IOWA. Do. do.
KANSAS. Do. do.
TENNESSEE. Do. do.
VIRGINIA. By State Teachers' Convention.
MISSISSIPPI. Do. do.
TEXAS. Do. do.

☞ This list includes nearly every State in which official recommendation is made.

New York City.	Louisville.
Brooklyn.	Newark.
New Orleans.	Milwaukee.
Buffalo.	Charleston.
Richmond.	Rochester.
Jersey City.	Mobile.
Hartford.	Syracuse.
Worcester.	Memphis.
Utica.	Savannah.
Wilmington.	Indianapolis.
Trenton.	Springfield.
Norfolk.	Wheeling.
Norwich.	Toledo.
Lockport.	Bridgeport.
Dubuque.	St. Paul.

And a multitude of less important points.

The Society of the CHRISTIAN BROTHERS, representing 40,000 pupils.
The FRANCISCAN BROTHERS, 8,000 pupils.
AMERICAN MISSIONARY SOCIETY, 50,000 pupils.

THE FREEDMEN,

By State Superintendents under the Freedmen's Bureau of

North Carolina. *Georgia.*
Louisiana. *Texas.*

For New triumphs of these truly *National* books, see current numbers of the Educational Bulletin,

MATHEMATICS.

DAVIES' NATIONAL COURSE.

ARITHMETIC.

SLATED.

1. Davies' Primary Arithmetic . . . . $ 25
2. Davies' Intellectual Arithmetic 45
3. Davies' Elements of Written Arithmetic . . . 50 $ 65
4. Davies' Practical Arithmetic 1 00   1 10
    Key to Practical Arithmetic *1 00
5. Davies' University Arithmetic. 1 40   1 55
    Key to University Arithmetic *1 40

ALGEBRA.

1. Davies' New Elementary Algebra 1 25   1 40
    Key to Elementary Algebra *1 25
2. Davies' University Algebra 1 60   1 75
    Key to University Algebra *1 60
3. Davies' Bourdon's Algebra 2 25   2 45
    Key to Bourdon's Algebra *2 25

GEOMETRY.

1. Davies' Elementary Geometry and Trigonometry 1 40   1 55
2. Davies' Legendre's Geometry 2 25   2 45
3. Davies' Analytical Geometry and Calculus . . 2 50   2 70
4. Davies' Descriptive Geometry 2 75   3 00

MENSURATION.

1. Davies' Practical Mathematics and Mensuration 1 40   1 55
2. Davies' Surveying and Navigation 2 50   2 70
3. Davies' Shades, Shadows, and Perspective . . 3 75   4 00

MATHEMATICAL SCIENCE.

Davies' Grammar of Arithmetic . . . .* 50
Davies' Outlines of Mathematical Science *1 00
Davies' Logic and Utility of Mathematics *1 50
Davies & Peck's Dictionary of Mathematics *3 50

DAVIES' NATIONAL COURSE of MATHEMATICS.

ITS RECORD.

In claiming for this series the first place among American text-books, of whatever class, the Publishers appeal to the magnificent record which its volumes have earned during the *thirty-five years* of Dr. Charles Davies' mathematical labors. The unremitting exertions of a life-time have placed *the modern series* on the same proud eminence among competitors that each of its predecessors has successively enjoyed in a course of constantly improved editions, now rounded to their perfect fruition—for it seems indeed that this science is susceptible of no further demonstration.

During the period alluded to, many authors and editors in this department have started into public notice, and by borrowing ideas and processes original with Dr. Davies, have enjoyed a brief popularity, but are now almost unknown. Many of the series of to-day, built upon a similar basis, and described as "modern books," are destined to a similar fate ; while the most far-seeing eye will find it difficult to fix the time, on the basis of any data afforded by their past history, when these books will cease to increase and prosper, and fix a still firmer hold on the affection of every educated American.

One cause of this unparalleled popularity is found in the fact that the enterprise of the author did not cease with the original completion of his books. Always a practical teacher, he has incorporated in his text-books from time to time the advantages of every improvement in methods of teaching, and every advance in science. During all the years in which he has been laboring, he constantly submitted his own theories and those of others to the practical test of the class-room—approving, rejecting, or modifying them as the experience thus obtained might suggest. In this way he has been able to produce an almost perfect series of class-books, in which every department of mathematics has received minute and exhaustive attention.

Nor has he yet retired from the field. Still in the prime of life, and enjoying a ripe experience which no other living mathematician or teacher can emulate, his pen is ever ready to carry on the good work, as the progress of science may demand. Witness his recent exposition of the " Metric System," which received the official endorsement of Congress, by its Committee on Uniform Weights and Measures.

DAVIES' SYSTEM IS THE ACKNOWLEDGED NATIONAL STANDARD FOR THE UNITED STATES, for the following reasons :—

1st. It is the basis of instruction in the great national schools at West Point and Annapolis.

2d. It has received the *quasi* endorsement of the National Congress.

3d. It is exclusively used in the public schools of the National Capital.

4th. The officials of the Government use it as authority in all cases involving mathematical questions.

5th. Our great soldiers and sailors commanding the national armies and navies were educated in this system. So have been a majority of eminent scientists in this country. All these refer to " Davies" as authority.

6th. A larger number of American citizens have received their education from this than from any other series.

7th. The series has a larger circulation throughout the whole country than any other, being *extensively used in every State in the Union.*

MATHEMATICS—Continued.

ARITHMETICAL EXAMPLES.

Reuck's Examples in Denominate Numbers $ 50

Reuck's Examples in Arithmetic · · · · · 1 00

These volumes differ from the ordinary arithmetic in their peculiarly *practical* character. They are composed mainly of examples, and afford the most severe and thorough discipline for the mind. While a book which should contain a complete treatise of theory and practice would be too cumbersome for every-day use, the insufficiency of *practical* examples has been a source of complaint.

HIGHER MATHEMATICS.

Church's Elements of Calculus · · · · · 2 50

Church's Analytical Geometry · · · · · · 2 50

Church's Descriptive Geometry, with Shades,

Shadows, and Perspective · · · · · · · 4 00

These volumes constitute the "West Point Course" in their several departments.

Courtenay's Elements of Calculus · · · · 3 00

A work especially popular at the South.

Hackley's Trigonometry · · · · · · · · · 3 00

With applications to navigation and surveying, nautical and practical geometry and geodesy, and logarithmic, trigonometrical, and nautical tables.

SLATED ARITHMETICS.

The Publishers have the pleasure to announce that they have perfected arrangements with the proprietor of Jocelyn's patent for Slated Books, whereby the "National Series of School Books" will enjoy the exclusive use of this remarkable and valuable invention It consists of the application of an artificially slated surface to the inner cover of a book, with flap of the same opening outward, so that students may refer to the book and use the slate at one and the same time, and as though the slate were detached. When folded up, the slate preserves examples and memoranda till needed. The material used is as durable as the stone slate. The additional cost of books thus improved is trifling.

THE METRIC SYSTEM.

Resolution of the Committee of the House of Representatives on a "Uniform System of Coinage, Weights, and Measures."

Be it Resolved, That Professor Charles Davies, LL.D., of the State of New York, be requested to confer with superintendents of public instruction, and teachers of schools, and others interested in a reform of the present incongruous system, and by lectures and addresses, to promote its general introduction and use.

The official version of the Metric System, as prepared by Dr. Davies, may be found in the Written, Practical, and University Arithmetics of the Mathematical Series, and is also published separately, price postpaid, *five cents,*

Davies' National Course of Mathematics.

TESTIMONIALS.

From L. Van Bokkelen, State Superintendent Public Instruction, Maryland.

The series of Arithmetics edited by Prof. Davies, and published by your firm, have been used for many years in the schools of several counties, and the city of Baltimore, and have been approved by teachers and commissioners.

Under the law of 1865, establishing a uniform system of Free Public Schools, these Arithmetics were unanimously adopted by the State Board of Education, after a careful examination, and are now used in all the Public Schools of Maryland.

These facts evidence the high opinion entertained by the School Authorities of the value of the series theoretically and practically.

From Horace Webster, President of the College of New York.

The undersigned has examined, with care and thought, several volumes of Davies' Mathematics, and is of the opinion that, as a whole, it is the most complete and best course for Academic and Collegiate instruction with which he is acquainted.

From David N. Camp, State Superintendent of Common Schools, Connecticut.

I have examined Davies' Series of Arithmetics with some care. The language is clear and precise; each principle is thoroughly analyzed, and the whole so arranged as to facilitate the work of instruction. Having observed the satisfaction and success with which the different books have been used by eminent teachers, it gives me pleasure to commend them to others.

From J. O. Wilson, Chairman Committee on Text-Books, Washington, D. C.

I consider Davies' Arithmetics decidedly superior to any other series, and in this opinion I am sustained, I believe, by the entire Board of Education and Corps of Teachers in this city, where they have been used for several years past.

From John L. Campbell, Professor of Mathematics, Wabash College, Indiana.

A proper combination of abstract reasoning and practical illustration is the chief excellence in Prof. Davies' Mathematical works. I prefer his Arithmetics, Algebras, Geometry, and Trigonometry to all others now in use, and cordially recommend them to all who desire the advancement of sound learning.

From Major J. H. Whittlesey, Government Inspector of Military Schools.

Be assured I regard the works of Professor Davies, with which I am acquainted, as by far the best text-books in print on the subjects which they treat. I shall certainly encourage their adoption wherever a word from me may be of any avail.

From T. McC. Ballantine, Professor Mathematics, Cumberland College, Kentucky.

I have long taught Prof. Davies' Course of Mathematics, and I continue to like their working.

From John McLean Bell, B. A., Principal of Lower Canada College.

I have used Davies' Arithmetical and Mathematical Series as text-books in the schools under my charge for the last six years. These I have found of great efficacy in exciting, invigorating, and concentrating the intellectual faculties of the young.

Each treatise serves as an introduction to the next higher, by the similarity of its reasonings and methods; and the student is carried forward, by easy and gradual steps, over the whole field of mathematical inquiry, and that, too, in a *shorter* time than is usually occupied in mastering a single department. I sincerely and heartily recommend them to the attention of my fellow-teachers in Canada.

From D. W. Steele, Prin. Philekolan Academy, Cold Springs, Texas.

I have used Davies' Arithmetics till I know them nearly by heart. A better series of school-books never were published. I have recommended them until they are now used in all this region of country.

A large mass of similar " Opinions" may be obtained by addressing the publishers for special circular for Davies' Mathematics. New recommendations are published in current numbers of the *Educational Bulletin.*

HISTORY.

Monteith's Youth's History, ... $ 70

A History of the United States for beginners. It is arranged upon the catechetical plan, with illustrative maps and engravings, review questions, dates in parentheses (that their study may be optional with the younger class of learners), and interesting Biographical Sketches of all persons who have been prominently identified with the history of our country.

Willard's United States, Sch. ed., $1 40. Un. ed. 2 25

Do.　　do.　　University edition, . 2 25

The plan of this standard work is chronologically exhibited in front of the title-page; the Maps and Sketches are found useful assistants to the memory, and dates, usually so difficult to remember, are so systematically arranged as in a great degree to obviate the difficulty. Candor, impartiality, and accuracy, are the distinguishing features of the narrative portion.

Willard's Universal History, ... 2 25

The most valuable features of the "United States" are reproduced in this. The peculiarities of the work are its great conciseness and the prominence given to the chronological order of events. The margin marks each successive era with great distinctness, so that the pupil retains not only the event but its time, and thus fixes the order of history firmly and usefully in his mind. Mrs. Willard's books are constantly revised, and at all times written up to embrace important historical events of recent date.

Berard's History of England, ... 1 75

By an authoress well known for the success of her History of the United States. The social life of the English people is felicitously interwoven, as in fact, with the civil and military transactions of the realm.

Ricord's History of Rome, ... 1 60

Possesses the charm of an attractive romance. The Fables with which this history abounds are introduced in such a way as not to deceive the inexperienced, while adding materially to the value of the work as a reliable index to the character and institutions, as well as the history of the Roman people.

Hanna's Bible History, ... 1 25

The only compendium of Bible narrative which affords a connected and chronological view of the important events there recorded, divested of all superfluous detail.

Summary of History, Complete 60

American History, $0 40. French and Eng. Hist. 35

A well proportioned outline of leading events, condensing the substance of the more extensive text-book in common use into a series of statements so brief, that every word may be committed to memory, and yet so comprehensive that it presents an accurate though general view of the whole continuous life of nations.

Marsh's Ecclesiastical History, ... 2 00

Questions to ditto, ... 75

Affording the History of the Church in all ages, with accounts of the pagan world during Biblical periods, and the character, rise, and progress of all Religions, as well as the various sects of the worshipers of Christ. The work is entirely non-sectarian, though strictly catholic.

PENMANSHIP.

Beers' System of Progressive Penmanship.
Per dozen $2 25

This "round hand" system of Penmanship in twelve numbers, commends itself by its simplicity and thoroughness. The first four numbers are primary books. Nos. 5 to 7, advanced books for boys. Nos. 8 to 10, advanced books for girls. Nos. 11 and 12, ornamental penmanship. These books are printed from steel plates (engraved by McLees), and are unexcelled in mechanical execution. Large quantities are annually sold.

Beers' Slated Copy Slips, per set *50

All beginners should practice, for a few weeks, slate exercises, familiarizing them with the form of the letters, the motions of the hand and arm, &c., &c. These copy slips, 32 in number, supply all the copies found in a complete series of writing-books, at a trifling cost.

Payson, Dunton & Scribner's Copy-B'ks. P. doz., 2 25

The National System of Penmanship, in three distinct series—(1) Common School Series, comprising the first six numbers: (2) Business Series, Nos. 8, 11, and 12; (3) Ladies' Series, Nos. 7, 9, and 10.

Fulton & Eastman's Chirographic Charts, *3 75

To embellish the school room walls, and furnish class exercise in the elements of Penmanship.

Payson's Copy-Book Cover, per hundred .*3 00

Protects every page except the one in use, and furnishes "lines" with proper slope for the penman, under. Patented.

National Steel Pens, Card with all kinds . . . *15

Pronounced by competent judges the perfection of American-made pens, and superior to any foreign article.

SCHOOL SERIES.		Index Pen, per gross . . . 75	
School Pen, per gross, . .$ 60		BUSINESS SERIES.	
Academic Pen, do . . 63		Albata Pen, per gross, . . 40	
Fine Pointed Pen, per gross 70		Bank Pen, do . . 70	
POPULAR SERIES.		Empire Pen, do . . 70	
Capitol Pen, per gross, . . 1 00		Commercial Pen, per gross . 60	
do do pr. box of 2 doz. 25		Express Pen do . 75	
Bullion Pen (imit. gold) pr. gr. 75		Falcon Pen, do . 70	
Ladies' Pen do 63		Elastic Pen, do . 75	

Stimpson's Scientific Steel Pen, per gross .*2 00

One forward and two backward arches, ensuring great strength, well-balanced elasticity, evenness of point, and smoothness of execution. One gross in twelve contains a Scientific Gold Pen.

Stimpson's Ink-Retaining Holder, per doz. .*2 00

A simple apparatus, which does not get out of order, withholds at a single dip as much ink as the pen would otherwise realize from a dozen trips to the inkstand, which it supplies with moderate and easy flow.

Stimpson's Gold Pen, $3 00; with Ink Retainer *4 50
Stimpson's Penman's Card, ...* 50

One dozen Steel Pens (assorted points) and Patent Ink-retaining Penholder.

BOOK-KEEPING.

Smith & Martin's Book-keeping · · · · $1 25
Blanks to ditto · · · · · · · · · · *60

This work is by a practical teacher and a practical book-keeper. It is of a thoroughly popular class, and will be welcomed by every one who loves to see theory and practice combined in an easy, concise, and methodical form.

The Single Entry portion is well adapted to supply a want felt in nearly all other treatises, which seem to be prepared mainly for the use of wholesale merchants, leaving retailers, mechanics, farmers, &c., who transact the greater portion of the business of the country, without a guide. The work is also commended, on this account, for general use in Young Ladies' Seminaries, where a thorough grounding in the simpler form of accounts will be invaluable to the future housekeepers of the nation.

The treatise on Double Entry Book-keeping combines all the advantages of the most recent methods, with the utmost simplicity of application, thus affording the pupil all the advantages of actual experience in the counting-house, and giving a clear comprehension of the entire subject through a judicious course of mercantile transactions.

The shape of the book is such that the transactions can be presented as in actual practice; and the simplified form of Blanks, three in number, adds greatly to the ease experienced in acquiring the science.

DRAWING.

The Little Artist's Portfolio · · · · · · · *50

25 Drawing Cards (progressive patterns), 25 Blanks, and a fine Artist's Pencil, all in one neat envelope.

Clark's Elements of Drawing · · · · · .*1 00

Containing full instructions, with appropriate designs and copies for a complete course in this graceful art, from the first rudiments of outline to the finished sketches of landscape and scenery.

Fowle's Linear and Perspective Drawing *60

For the cultivation of the eye and hand, with copious illustrations and directions, which will enable the unskilled teacher to learn the art himself while instructing his pupils.

Monk's Drawing Books—Six Numbers, per set*2 25

A series of progressive Drawing Books, presenting copy and blank on opposite pages. The copies are fac-similes of the best imported lithographs, the originals of which cost from 50 cents to $1.50 *each* in the print-stores. Each book contains *eleven* large patterns. No. 1.—Elementary studies; No. 2.—Studies of Foliage; No. 3.—Landscapes; No. 4.—Animals, I.; No. 5.—Animals, II.; No. 6.—Marine Views, &c.

Ripley's Map Drawing · · · · · · · · · .1 25

One of the most efficient aids to the acquirement of a knowledge of geography is the practice of map drawing. It is useful for the same reason that the best exercise in orthography is the *writing* of difficult words. Sight comes to the aid of hearing, and a double impression is produced upon the memory. Knowledge becomes less mechanical and more intuitive. The student who has sketched the outlines of a country, and dotted the important places, is little likely to forget either. The impression produced may be compared to that of a traveler who has been over the ground, while more comprehensive and accurate in detail.

21

NATURAL SCIENCE.

FAMILIAR SCIENCE

Norton & Porter's First Book of Science, · $1 75

By eminent Professors of Yale College. Contains the principles of Natural Philosophy, Astronomy, Chemistry, Physiology, and Geology. Arranged on the Catechetical plan for primary classes and beginners.

Chambers' Treasury of Knowledge, · · · 1 25

Progressive lessons upon—*first*, common things which lie most immediately around us, and first attract the attention of the young mind; *second*, common objects from the Mineral, Animal, and Vegetable kingdoms, manufactured articles, and miscellaneous substances; *third*, a systematic view of Nature under the various sciences. May be used as a Reader or Text-Book.

NATURAL PHILOSOPHY.

Norton's First Book in Natural Philosophy, 1 00

By Prof. NORTON, of Yale College. Designed for beginners; profusely illustrated, and arranged on the Catechetical plan.

Peck's Ganot's Course of Nat. Philosophy, 1 75

The standard text-book of France, Americanized and popularized by Prof. PECK, of Columbia College. The most magnificent system of illustration ever adopted in an American school-book is here found. For intermediate classes.

Peck's Elements of Mechanics, · · · · · · 2 25

A suitable introduction to Bartlett's higher treatises on Mechanical Philosophy, and adequate in itself for a complete academical course.

Bartlett's Synthetic Mechanics, · · · · · 3 75
Bartlett's Analytical Mechanics, · · · · · 5 50
Bartlett's Acoustics and Optics, · · · · · 3 0C

A system of Collegiate Philosophy, by Prof. BARTLETT, of West Point Military Academy.

Steele's 14 Weeks Course in Philosophy, · 1 50

GEOLOGY.

Page's Elements of Geology, · · · · · · · 1 25

A volume of Chambers' Educational Course. Practical, simple, and eminently calculated to make the study interesting.

Emmon's Manual of Geology, · · · · · · 1 25

The first Geologist of the country has here produced a work worthy of his reputation. The plan of presenting the subject is an obvious improvement on older methods. The department of Palæontology receives especial attention.

22

Peck's Ganot's Popular Physics.

TESTIMONIALS.

From PROF. ALONZO COLLIN, *Cornell College, Iowa.*

I am pleased with it. I have decided to introduce it as a text-book.

From II. F. JOHNSON, *President Madison College, Sharon, Miss.*

I am pleased with Peck's Ganot, and think it a magnificent book.

From PROF. EDWARD BROOKS, *Pennsylvania State Normal School*

So eminent are its merits, that it will be introduced as the text-book upon elementary physics in this institution.

From II. II. LOCKWOOD, *Professor Natural Philosophy U. S. Naval Academy.*

I am so pleased with it that I will probably add it to a course of lectures given to the midshipmen of this school on physics.

From GEO. S. MACKIE, *Professor Natural History University of Nashville, Tenn.*

I have decided on the introduction of Peck's Ganot's Philosophy, as I am satisfied that it is the best book for the purposes of my pupils that I have seen, combining simplicity of explanation with elegance of illustration.

From W. S. McRAE, *Superintendent Vevay Public Schools, Indiana.*

Having carefully examined a number of text-books on natural philosophy, I do not hesitate to express my decided opinion in favor of Peck's Ganot. The matter, style, and illustration eminently adapt the work to the popular wants.

From REV. SAMUEL McKINNEY, D.D., *President Austin College, Huntsville, Texas.*

It gives me pleasure to commend it to teachers. I have taught some classes with it as our text, and must say, for simplicity of style and clearness of illustration, I have found nothing as yet published of equal value to the teacher and pupil.

From C. V. SPEAR, *Principal Maplewood Institute, Pittsfield, Mass.*

I am much pleased with its ample illustrations by plates, and its clearness and simplicity of statement. It covers the ground usually gone over by our higher classes, and contains many fresh illustrations from life or daily occurrences and new applications of scientific principles to such.

From J. A. BANFIELD, *Superintendent Marshall Public Schools, Michigan.*

I have used Peck's Ganot since 1863, and with increasing pleasure and satisfaction each term. I consider it superior to any other work on physics in its adaptation to our high schools and academies. Its illustrations are superb—better than three times their number of pages of fine print.

From A. SCHUYLER, *Professor of Mathematics in Baldwin University, Berea, Ohio.*

After a careful examination of Peck's Ganot's Natural Philosophy, and an actual test of its merits as a text-book, I can heartily recommend it as admirably adapted to meet the wants of the grade of students for which it intended. Its diagrams and illustrations are *unrivaled.* We use it in the Baldwin University.

From D. C. VAN NORMAN, *Principal Van Norman Institute, New York.*

The Natural Philosophy of M. Ganot, edited by Prof. Peck, is, in my opinion, the best work of its kind, for the use intended, ever published in this country. Whether regarded in relation to the natural order of the topics, the precision and clearness of its definitions, or the fullness and beauty of its illustrations, it is certainly, I think, an advance.

☞ For many similar testimonials, see current numbers of the Illustrated Educational Bulletin.

NATURAL SCIENCE—Continued.

CHEMISTRY.

Porter's First Book of Chemistry, · · · .$1 00

Porter's Principles of Chemistry, · · · · 2 00

The above are widely known as the productions of one of the most eminent scientific men of America. The extreme simplicity in the method of presenting the science, while exhaustively treated, has excited universal commendation. Apparatus adequate to the performance of every experiment mentioned, may be had of the publishers for a trifling sum. The effort to popularize the science is a great success. It is now within the reach of the poorest and least capable at once.

Darby's Text-Book of Chemistry, · · · · 1 75

Purely a Chemistry, divesting the subject of matters comparatively foreign to it (such as heat, light, electricity, etc.), but usually allowed to engross too much attention in ordinary school-books.

Gregory's Organic Chemistry, · · · · · 2 50

Gregory's Inorganic Chemistry, · · · · · 2 50

The science exhaustively treated. For colleges and medical students.

Steele's Fourteen Weeks' Course, · · · · · 1 25

A successful effort to reduce the study to the limits of a *single term*, thereby making feasible its general introduction in institutions of every character. The author's felicity of style and success in making the science pre-eminently *interesting* are peculiarly noticeable features.

Chemical Apparatus, to accompany "Porter"   20 00
    do   do   to accompany "Steele"   25 00

BOTANY.

Thinker's First Lessons in Botany, · · · ·   40

For children. The technical terms are largely dispensed with in favor of an easy and familiar style adapted to the smallest learner.

Wood's Object Lessons in Botany, · · · · 1 35

Wood's Intermediate Botany, · · · · · 2 25

Wood's New Class-Book of Botany, · · · 3 50

The standard text-books of the United States in this department. In style they are simple, popular, and lively; in arrangement, easy and natural; in description, graphic and strictly exact. The Tables for Analysis are reduced to a perfect system. More are annually sold than of all others combined.

Darby's Southern Botany, · · · · · · · 2 00

Embracing general Structural and Physiological Botany, with vegetable products, and descriptions of Southern plants, and a complete Flora of the Southern States.

24

PHYSIOLOGY.

Jarvis' Primary Physiology, ...$ 75
Jarvis' Physiology and Laws of Health, . 1 65

The only books extant which approach this subject with a proper view of the true object of teaching Physiology in schools, viz., that scholars may know how to take care of their own health. In bold contrast with the abstract *Anatomies*, which children learn as they would Greek or Latin (and forget as soon), to *discipline the mind*, are these text-books, using the *science* as a secondary consideration, and only so far as is necessary for the comprehension of the *laws of health.*

Hamilton's Vegetable & Animal Physiology, 1 25

The two branches of the science combined in one volume lead the student to a proper comprehension of the Analogies of Nature.

ASTRONOMY.

Willard's School Astronomy, ...1 00

By means of clear and attractive illustrations, addressing the eye in many cases by analogies, careful definitions of all necessary technical terms, a careful avoidance of verbiage and unimportant matter, particular attention to analysis, and a general adoption of the simplest methods, Mrs. Willard has made the best and most attractive *elementary* Astronomy extant.

McIntyre's Astronomy and the Globes, . . 1 50

A complete treatise for intermediate classes. Highly approved.

Bartlett's Spherical Astronomy, ... 4 50

The West Point course, for advanced classes, with applications to the current wants of Navigation, Geography, and Chronology.

NATURAL HISTORY.

Carl's Child's Book of Natural History, . . 0 50

Illustrating the Animal, Vegetable, and Mineral Kingdoms, with application to the Arts. For beginners. Beautifully and copiously illustrated.

ZOOLOGY.

Chambers' Elements of Zoology, ... 1 50

A complete and comprehensive system of Zoology, adapted for academic instruction, presenting a systematic view of the Animal Kingdom as a portion of external Nature.

It will be observed, that, in the various departments of Natural Science, the NATIONAL SERIES is extremely rich. The mineral, animal, and vegetable kingdoms, matter, and the laws that govern it in all its forms, are here placed before the student by those who have made its study a specialty and a life work. The works of Professors PECK, of Columbia College, NORTON AND PORTER, of Yale, BARTLETT, of West Point Military Academy, EMMONS, of Williams, and State Geologist of New York and North Carolina, WOOD, the botanist, and JARVIS, the eminent physician, are esteemed indubitable authority in all that concerns their several specialties

Jarvis' Physiology and Laws of Health.

TESTIMONIALS.

From SAMUEL B. McLANE, *Superintendent Public Schools, Keokuk, Iowa.*

I am glad to see a really good text-book on this much neglected branch. This is clear, concise, accurate, and eminently adapted to the *class-room.*

From WILLIAM F. WYERS, *Principal of Academy, West Chester, Pennsylvania.*

A thorough examination has satisfied me of its superior claims as a text-book to the attention of teacher and taught. I shall introduce it at once.

From H. R. SANFORD, *Principal of East Genesee Conference Seminary, N. Y.*

"Jarvis' Physiology" is received, and fully met our expectations. We immediately adopted it.

From ISAAC T. GOODNOW, *State Superintendent of Kansas—published in connection with the "School Law."*

"Jarvis' Physiology," a common-sense, practical work, with just enough of anatomy to understand the physiological portions. The last six pages, on Man's Responsibility for his own health, are worth the price of the book.

From D. W. STEVENS, *Superintendent Public Schools, Fall River, Mass.*

I have examined Jarvis' "Physiology and Laws of Health," which you had the kindness to send to me a short time ago. In my judgment it is far the best work of the kind within my knowledge. It has been adopted as a text-book in our public schools.

From HENRY G. DENNY, *Chairman Book Committee, Boston, Mass.*

The very excellent "Physiology" of Dr. Jarvis I had introduced into our High School, where the study had been temporarily dropped, believing it to be by far the best work of the kind that had come under my observation; indeed, the reintroduction of the study was delayed for some months, because Dr. Jarvis' book could not be had, and we were unwilling to take any other.

From PROF. A. P. PEABODY, D.D., LL.D., *Harvard University.*

 * * I have been in the habit of examining school-books with great care, and I hesitate not to say that, of all the text-books on Physiology which have been given to the public, Dr. Jarvis' deserves the first place on the score of accuracy, thoroughness, method, simplicity of statement, and constant reference to topics of practical interest and utility.

From JAMES N. TOWNSEND, *Superintendent Public Schools, Hudson, N. Y.*

Every human being is appointed to take charge of his own body; and of all books written upon this subject, I know of none which will so well prepare one to do this as " Jarvis' Physiology "—that is, in so small a compass of matter. It considers the pure, simple *laws of health* paramount to science; and though the work is thoroughly scientific, it is divested of all cumbrous technicalities, and presents the subject of physical life in a manner and style really charming. It is unquestionably the best text-book on physiology I have ever seen. It is giving great satisfaction in the schools of this city, where it has been adopted as the standard.

From L. J. SANFORD, M.D., *Prof. Anatomy and Physiology in Yale College.*

Books on human physiology, designed for the use of schools, are more generally a failure perhaps than are school-books on most other subjects.

The great want in this department is met, we think, in the well-written treatise of Dr. Jarvis, entitled "Physiology and Laws of Health." * * The work is not too detailed nor too expansive in any department, and is clear and concise in all. It is not burdened with an excess of anatomical description, nor rendered discursive by many zoological references. Anatomical statements are made to the extent of qualifying the student to attend, understandingly, to an exposition of those functional processes which, collectively, make up health; thus the laws of health are enunciated, and many suggestions are given which, if heeded, will tend to its preservation.

☞ For further testimony of similar character, see current numbers of the Illustrated Educational Bulletin.

MODERN LANGUAGE.

French and English Primer, ...$ 10
German and English Primer, ... 10
Spanish and English Primer, ... 10
The names of common objects properly illustrated and arranged in easy lessons.

Ledru's French Fables, ... 75
Ledru's French Grammar, ... 1 00
Ledru's French Reader, ... 1 00
The author's long experience has enabled him to present the most thoroughly practical text-books extant, in this branch. The system of pronunciation (by phonetic illustration) is original with this author, and will commend itself to all American teachers, as it enables their pupils to secure an absolutely correct pronunciation without the assistance of a native master. This feature is peculiarly valuable also to "self-taught" students. The directions for ascertaining the gender of French nouns—also a great stumbling-block—are peculiar to this work, and will be found remarkably competent to the end proposed. The criticism of teachers and the test of the school-room is invited to this excellent series, with confidence.

Haskin's French and English First Book . 75
Presents the striking feature of a simultaneous presentation of the elementary principles of the vernacular with those of a foreign language. This is the method which the practical teacher naturally pursues in oral instruction, and possesses peculiar advantages in application to young pupils.

Pujol's Complete French Class-Book, . . . 2 25
Offers, in one volume, methodically arranged, a complete French course—usually embraced in series of from five to twelve books, including the bulky and expensive Lexicon. Here are Grammar, Conversation, and choice Literature—selected from the best French authors. Each branch is thoroughly handled; and the student, having diligently completed the course as prescribed, may consider himself, without further application, *au fait* in the most polite and elegant language of modern times.

Maurice-Poitevin's Grammaire Francaise, . 1 00
American schools are at last supplied with an American edition of this famous text-book. Many of our best institutions have for years been procuring it from abroad rather than forego the advantages it offers. The policy of putting students who have acquired some proficiency from the ordinary text-books, into a Grammar written in the vernacular, can not be too highly commended. It affords an opportunity for finish and review at once; while embodying abundant practice of its own rules.

Worman's Elementary German Grammar, . 1 2
A work of great merit. Well calculated to ground the student in the elements of this language, become so important by the extensive settlement of Germans in this country.

Willard's Historia de los Estados Unidos, . 2 00
The History of the United States, translated by Professors TOLON and DE TORNOS, will be found a valuable, instructive, and entertaining reading-book for Spanish classes.

Pujol's Complete French Class-Book.

' TESTIMONIALS.'

From Prof. Elias Peissner, *Union College.*

I take great pleasure in recommending Pujol and Van Norman's French Class-Book, as there is no French grammar or class-book which can be compared with it in completeness, system, clearness, and general utility.

From Edward North, *President of Hamilton College.*

I have carefully examined Pujol and Van Norman's French Class-Book, and am satisfied of its superiority, for college purposes, over any other heretofore used. We shall not fail to use it with our next class in French.

From A. Curtis, *Pres't of Cincinnati Literary and Scientific Institute.*

I am confident that it may be made an instrument in conveying to the student, in from six months to a year, the art of speaking and writing the French with almost native fluency and propriety.

From Hiram Orcutt, A. M., *Prin. Glenwood and Tilden Ladies' Seminaries.*

I have used Pujol's French Grammar in my two seminaries, exclusively, for more than a year, and have no hesitation in saying that I regard it the best text-book in this department extant. And my opinion is confirmed by the testimony of Prof. F. De Launay and Mademoiselle Marindin. They assure me that the book is eminently accurate and practical, as tested in the school-room.

From Prof. Theo. F. De Fumat, *Hebrew Educational Institute, Memphis, Tenn.*

M. Pujol's French Grammar is one of the best and most practical works. The French language is chosen and elegant in style—modern and easy. It is far superior to the other French class-books in this country. The selection of the conversational part is very good, and will interest pupils; and being all completed in only one volume, it is especially desirable to have it introduced in our schools.

From Prof. James H. Worman, *Bordentown Female College, N. J.*

The work is upon the same plan as the text-books for the study of French and English published in Berlin, for the study of those who have not the aid of a teacher, and these books are considered, by the first authorities, the best books. In most of our institutions, Americans teach the modern languages, and heretofore the trouble has been to give them a text-book that would dispose of the difficulties of the French pronunciation. This difficulty is successfully removed by P. and Van N., and I have every reason to believe it will soon make its way into most of our best schools.

From Prof. Charles S. Dod, *Ann Smith Academy, Lexington, Va.*

I cannot do better than to recommend "Pujol and Van Norman." For comprehensive and systematic arrangement, progressive and thorough development of all grammatical principles and idioms, with a due admixture of theoretical knowledge and practical exercise, I regard it as superior to any (other) book of the kind.

From A. A. Forster, *Prin. Pinehurst School, Toronto, C. W.*

I have great satisfaction in bearing testimony to M. Pujol's System of French Instruction, as given in his complete class-book. For clearness and comprehensiveness, adapted for all classes of pupils, I have found it superior to any other work of the kind, and have now used it for some years in my establishment with great success.

From Prof. Otto Fedder, *Maplewood Institute, Pittsfield, Mass.*

The conversational exercises will prove an immense saving of the hardest kind of labor to teachers. There is scarcely any thing more trying in the way of teaching language, than to rack your brain for short and easily intelligible bits of conversation, and to repeat them time and again with no better result than extorting at long intervals a doubting " oui," or a hesitating " non, monsieur."

☞ For further testimony of a similar character, see special circular, and current numbers of the Educational Bulletin.

Worman's German Grammars.

TESTIMONIALS.

From Sec'y Chas. A. Eggert, *Iowa State University,*

A work of superior merit. As a text-book it is *the best* I have ever seen.

From Prof. R. W. Jones, *Petersburg Female College, Va.*

From what I have seen of the work it is almost certain *I shall introduce it* into this institution.

From Prof. G. Campbell, *University of Minnesota.*

A valuable addition to our school-books, and will find many friends, and do great good.

From Prof. O. H P. Corbeew, *Mary Military Inst., Md.*

I am better pleased with them than any I have ever taught. I have already ordered through our booksellers.

From Prof. R. S. Kendall, *Vernon Academy, Conn.*

I at once put the Elementary Grammar into the hands of a class of beginners, and have used it *with great satisfaction.*

From Prof. D. E. Holmes, *Berlin Academy, Wis.*

Worman's German works are *superior.* I shall use them hereafter in my German classes.

From Prof. Magnus Buchholtz, *Hiram College, Ohio.*

I have examined the Complete Grammar, and find it *excellent.* You may rely that it will be used here.

From Prin. Thos. W. Tobey, *Paducah Female Seminary, Ky.*

The Complete German Grammar is worthy of an extensive circulation. It is *admirably adapted* to the class-room. I shall use it.

From Prof. Alex. Rosenspitz, *Houston Academy, Texas.*

Bearer will take and pay for 3 dozen copies. Mr. Worman deserves the approbation and esteem of the teacher and the thanks of the student.

From Prof. G. Malmene, *Augusta Seminary, Maine.*

The Complete Grammar cannot fail to *give great satisfaction* by the simplicity of its arrangement, and by its completeness.

From Prin. Oval Pirkey, *Christian University, Mo.*

Just such a series as is positively necessary. I do hope the author will succeed as well in the French, &c., as he has in the German.

From Prof. S. D. Hillman, *Dickinson College, Pa.*

The class have lately commenced, and my examination thus far warrants me in saying that I regard it *as the best* grammar for instruction in the German.

From Prin. Silas Livermore, *Bloomfield Seminary, Mo.*

I have found a classically and scientifically educated Prussian gentleman whom I propose to make German instructor. I have shown him both your German grammars. He has expressed *his approbation* of them generally.

From Prof. Z. Test, *Howland School for Young Ladies, N. Y.*

I shall introduce the books. From a cursory examination I have no hesitation in pronouncing the Complete Grammar *a decided improvement* on the text-books at present in use in this country.

From Prof. Lewis Kistler, *Northwestern University, Ill.*

Having looked through the Complete Grammar with some care I must say that you have produced *a good book;* you may be awarded with this gratification—that your grammar promotes the facility of learning the German language, and of becoming acquainted with its rich literature.

From Pres. J. P. Rous, *Stockwell Collegiate Inst., Ind.*

I supplied a class with the Elementary Grammar, and it gives *complete satisfaction.* The conversational and reading exercises are well calculated to illustrate the principles, and lead the student on an easy yet thorough course. I think the Complete Grammar equally attractive.

THE CLASSICS.

LATIN.

Silber's Latin Course, ... $1 25

The book contains an Epitome of Latin Grammar, followed by Reading Exercises, with explanatory Notes and copious References to the leading Latin Grammars, and also to the Epitome which precedes the work. Then follow a Latin-English Vocabulary and Exercises in Latin Prose Composition, being thus complete in itself, and a very suitable work to put in the hands of one about to study the language.

Searing's Virgil's Æneid, ... 2 25

It contains only the first six books of the Æneid. 2. A very carefully constructed Dictionary. 3. Sufficiently copious Notes. 4. Grammatical references to four leading Grammars. 5. Numerous Illustrations of the highest order. 6. A superb Map of the Mediterranean and adjacent countries. 7. Dr. S. H. Taylor's "Questions on the Æneid." 8. A Metrical Index, and an Essay on the Poetical Style. 9. A photographic *fac simile* of an early Latin M.S. 10. The text according to Jahn, but paragraphed according to Ladewig. 11. Superior mechanical execution. 12. The price no greater than that of ordinary editions.

Hanson's Latin Prose Book, . . · 3 12
Hanson's Latin Poetry, ... 3 12

Andrews & Stoddard's Latin Grammar, *1 50
Andrews' Questions on the Grammar, . *0 15
Andrews' Latin Exercises, ... *1 25
Andrews' Viri Romæ, · *1 25
Andrews' Sallust's Jugurthine War, &c. *1 50
Andrews' Eclogues & Georgics of Virgil, *1 50
Andrews' Cæsar's Commentaries, ... *1 50
Andrews' Ovid's Metamorphoses, . . . *1 25

GREEK.

Crosby's Greek Grammar, ... 1 88
Crosby's Xenophon's Anabasis, ... 1 20

MYTHOLOGY.

Dwight's Grecian and Roman Mythology.

School edition, $1 25;   University edition, *2 25

A knowledge of the fables of antiquity, thus presented in a systematic form, is as indispensable to the student of general literature as to him who would peruse intelligently the classical authors. The mythological allusions so frequent in literature are readily understood with such a Key as this.

ELOCUTION.

Watson's Practical Elocution · · · · · · .$0 25
A brief, clear, and most satisfactory treatise—same as in " Independent Fifth Reader." The subject fully illustrated by diagrams.

Zachos' Analytic Elocution · · · · · · .1 50
All departments of elocution—such as the analysis of the voice and the sentence, phonology, rhythm, expression, gesture, &c.—are here arranged for instruction in classes, illustrated by copious examples.

Sherwood's Self Culture · · · · · · · · .1 00
Self-culture in reading, speaking, and conversation—a very valuable treatise to those who would perfect themselves in these accomplishments.

SPEAKERS.

Northend's Little Orator · · · · · · · · *60
Contains simple and attractive pieces in prose and poetry, adapted to the capacity of children under twelve years of age.

Northend's National Orator · · · · · · .*1 25
About one hundred and seventy choice pieces happily arranged. The design of the author in making the selection has been to cultivate *versatility of expression.*

Northend's Entertaining Dialogues · · · .*1 25
Extracts eminently adapted to cultivate the dramatic faculties, as well as entertain an audience.

Swett's Common School Speaker · · · .*1 25
Selections from recent literature.

Raymond's Patriotic Speaker · · · · · .*2 00
A superb compilation of modern eloquence and poetry, with original dramatic exercises. Nearly every eminent *living* orator is represented, without distinction of place or party.

COMPOSITION, &c.

Brookfield's First Book in Composition · 50
Making the cultivation of this important art feasible for the smallest child. By a new method, to induce and stimulate thought.

Boyd's Composition and Rhetoric · · · · 1 35
This work furnishes all the aid that is needful or can be desired in the various departments and styles of composition, both in prose and verse.

Day's Art of Rhetoric · · · · · · · · · 1 25
Noted for exactness of definition, clear limitation, and philosophical development of subject; the large share of attention given to Invention, as a branch of Rhetoric, and the unequalled analysis of style.

31

LITERATURE.

Boyd's Milton's Paradise Lost *1 25
Boyd's Young's Night Thoughts *1 25
Boyd's Cowper's Task, Table Talk, &c. .*1 25
Boyd's Thomson's Seasons *1 25
Boyd's Pollok's Course of Time *1 25
Boyd's Lord Bacon's Essays *1 60

> This series of annotated editions of great English writers, in prose and poetry, is designed for critical reading and parsing in schools. Prof. J. R. Boyd proves himself an editor of high capacity, and the works themselves need no encomium. As auxiliary to the study of Belles Lettres, etc., these works have no equal.

Pope's Essay on Man *20
Pope's Homer's Iliad *80

> The metrical translation of the great poet of antiquity, and the matchless "Essay on the Nature and State of Man," by ALEXANDER POPE, afford superior exercise in literature and parsing.

AESTHETICS.

Huntington's Manual of the Fine Arts . .*1 75

> A view of the rise and progress of Art in different countries, a brief account of the most eminent masters of Art, and an analysis of the principles of Art. It is complete in itself, or may precede to advantage the critical work of Lord Kames.

Boyd's Kames' Elements of Criticism . .*1 75

> The best edition of this standard work; without the study of which none may be considered proficient in the science of the Perceptions. No other study can be pursued with so marked an effect upon the taste and refinement of the pupil.

POLITICAL ECONOMY.

Champlin's Lessons on Political Economy   1 25

> An improvement on previous treatises, being shorter, yet containing every thing essential, with a view of recent questions in finance, etc., which is not elsewhere found.

MENTAL PHILOSOPHY.

Mahan's Intellectual Philosophy $1 75
The subject exhaustively considered. The author has evinced learning, candor, and independent thinking.

Mahan's Science of Logic 2 00
A profound analysis of the laws of thought. The system possesses the merit of being intelligible and self consistent. In addition to the author's carefully elaborated views, it embraces results attained by the ablest minds of Great Britain, Germany, and France, in this department.

Boyd's Elements of Logic 1 00
A systematic and philosophic condensation of the subject, fortified with additions from Watts, Abercrombie, Whately, &c.

Watts on the Mind 45
The Improvement of the Mind, by Isaac Watts, is designed as a guide for the attainment of useful knowledge. As a text-book it is unparalleled; and the discipline it affords cannot be too highly esteemed by the educator.

MORALS.

Alden's Text-Book of Ethics 60
For young pupils. To aid in systematizing the ethical teachings of the Bible, and point out the coincidences between the instructions of the sacred volume and the sound conclusions of reason.

Willard's Morals for the Young *75
Lessons in conversational style to inculcate the elements of moral philosophy. The study is made attractive by narratives and engravings.

GOVERNMENT.

Howe's Young Citizen's Catechism 60
Explaining the duties of District, Town, City, County, State, and United States Officers, with rules for parliamentary and commercial business—that which every future "sovereign" ought to know, and so few are taught.

Young's Lessons in Civil Government . . 1 25
A comprehensive view of Government, and abstract of the laws showing the rights, duties, and responsibilities of citizens.

Mansfield's Political Manual 1 25
This is a complete view of the theory and practice of the General and State Governments of the United States, designed as a text-book. The author is an esteemed and able professor of constitutional law, widely known for his sagacious utterances in matters of statecraft through the public press. Recent events teach with emphasis the vital necessity that the rising generation should comprehend the noble polity of the American government, that they may act intelligently when endowed with a voice in it.

TEACHERS' AIDS.

Brooks' School Manual of Devotion . . . 70

This volume contains daily devotional exercises, consisting of a hymn, selections of scripture for alternate reading by teacher and pupils, and a prayer. Its value for opening and closing school is apparent.

Cleaveland's School Harmonist *70

Contains appropriate *tunes* for each hymn in the "Manual of Devotion" described above.

The Boy Soldier 75

Complete infantry tactics for schools, with illustrations, for the use of those who would introduce this pleasing relaxation from the confining duties of the desk.

Welch's Object Lessons 1 00

Invaluable for teachers of primary schools. Contains the best explanation of the Pestalozzian system. By its aid the proficiency of pupils and the general interest of the school may be increased one hundred per cent.

Tracy's School Record *75

To record attendance, deportment, and scholarship; containing also many useful tables and suggestions to teachers, that are worth of themselves the price of the book.

Tracy's Pocket Record *65

A portable edition of the School Record, without the tables, &c.

Brooks' Teacher's Register *1 00

Presents at one view a record of attendance, recitations, and deportment for the whole term.

Carter's Record and Roll-Book . . . . .*2 50.

For large graded schools.

National School Diary, per dozen *1 00

A little book of blank forms for weekly report of the standing of each scholar, from teacher to parent. A great convenience.

THE
TEACHER'S LIBRARY.

The Student; or, Fireside Friend — Phelps,$*1.50

The Educator; or, Hours with my
Pupils, ... do., *1 50

The Discipline of Life; or, Ida
Norman,. . . . . do., *1 75

The authoress of these works is one of the most distinguished writers on education; and they can not fail to prove a valuable addition to the School and Teachers' Libraries, being in a high degree both interesting and instructive.

Hecker's Scientific Basis of Education, . .*2 5(
Adaptation of study and classification by temperaments.

Object Lessons—Welch *100
This is a complete exposition of the popular modern system of "object-teaching," for teachers of primary classes.

Theory and Practice of Teaching—Page .*1 50
This volume has, without doubt, been read by two hundred thousand teachers, and its popularity remains undiminished—large editions being exhausted yearly. It was the pioneer, as it is now the patriarch of professional works for teachers.

The Graded School—Wells *1 25
The proper way to organize graded schools is here illustrated. The author has availed himself of the best elements of the several systems prevalent in Boston, New York, Philadelphia, Cincinnati, St. Louis, and other cities.

The Normal—Holbrook *1 75
Carries a working school on its visit to teachers, showing the most approved methods of teaching all the common branches, including the technicalities, explanations, demonstrations, and definitions introductory and peculiar to each branch.

The Teachers' Institute—Fowle *1 25
This is a volume of suggestions inspired by the author's experience at institutes, in the instruction of young teachers. A thousand points of interest to this class are most satisfactorily dealt with.

The Teacher and the Parent—Northend . $*1 50

A treatise upon common-school education, designed to lead teachers to view their calling in its true light, and to stimulate them to fidelity.

The Teachers' Assistant—Northend . . .*1 50

A natural continuation of the author's previous work, more directly calculated for daily use in the administration of school discipline and instruction.

School Government—Jewell *1 50

Full of advanced ideas on the subject which its title indicates. The criticisms upon current theories of punishment and schemes of administration have excited general attention and comment.

Grammatical Diagrams—Jewell *1 00

The diagram system of teaching grammar explained, defended, and improved. The curious in literature, the searcher for truth, those interested in new inventions, as well as the disciples of Prof. Clark, who would see their favorite theory fairly treated, all want this book. There are many who would like to be made familiar with this system before risking its use in a class. The opportunity is here afforded.

The Complete Examiner—Stone *1 25

Consists of a series of questions on every English branch of school and academic instruction, with reference to a given page or article of leading text-books where the answer may be found in full. Prepared to aid teachers in securing certificates, pupils in preparing for promotion, and teachers in selecting review questions.

School Amusements—Root *1 50

To assist teachers in making the school interesting, with hints upon the management of the school-room. Rules for military and gymnastic exercises are included. Illustrated by diagrams.

Institute Lectures on Mental and Moral Culture—Bates *1 50

These lectures, originally delivered before institutes, are based upon various topics of interest to the teacher. The volume is calculated to prepare the will, awaken the inquiry, and stimulate the thought of the zealous teacher.

Method of Teachers' Institutes—Bates . . .* 75

Sets forth the best method of conducting institutes, with a detailed account of the object, organization, plan of instruction, and true theory of education on which such instruction should be based.

History and Progress of Education . . .*1 50

The systems of education prevailing in all nations and ages, the gradual advance to the present time, and the bearing of the past upon the present in this regard, are worthy of the careful investigation of all concerned in the cause.

American Education—Mansfield $1 50

A treatise on the principles and elements of education, as practiced in this country, with ideas towards distinctive republican and Christian education.

American Institutions—De Tocqueville . .*1 50

A valuable index to the genius of our Government.

Universal Education—Mayhew *1 75

The subject is approached with the clear, keen perception of one who has observed its necessity, and realized its feasibility and expediency alike. The redeeming and elevating power of improved common schools constitutes the inspiration of the volume.

Higher Christian Education—Dwight . . .*1 50

A treatise on the principles and spirit, the modes, directions, and results of all true teaching; showing that right education should appeal to every element of enthusiasm in the teacher's nature.

Modern Philology—Dwight *1 75

Important to the grammarian, and indispensable to the teacher of language, ancient or modern, who would afford his pupils the advantage of the analogy and association to be derived from an intelligent comparison of all languages and their history.

Lectures on Natural History—Chadbourne .* 75

Affording many themes for oral instruction in this interesting science—especially in schools where it is not pursued as a class exercise.

Outlines of Mathematical Science—Davies *1 00

A manual suggesting the best methods of presenting mathematical instruction on the part of the teacher, with that comprehensive view of the whole which is necessary to the intelligent treatment of a part, in science.

Logic & Utility of Mathematics—Davies . .*1 50

An elaborate and lucid exposition of the principles which lie at the foundation of pure mathematics, with a highly ingenious application of their results to the development of the essential idea of the different branches of the science.

Mathematical Dictionary—Davies & Peck . 3 50

This cyclopædia of mathematical science defines with completeness, precision, and accuracy, every technical term, thus constituting a popular treatise on each branch, and a general view of the whole subject.

School Architecture—Barnard 2 25

Attention is here called to the vital connection between a good schoolhouse and a good school, with plans and specifications for securing the former in the most economical and satisfactory manner.

THE SCHOOL LIBRARY.

The two elements of instruction and entertainment were never more happily combined than in this collection of standard books. Children and adults alike will here find ample food for the mind, of the sort that is easily *digested*, while not degenerating to the level of modern romance.

LIBRARY OF LITERATURE.

Milton's Paradise Lost   Boyd's Illustrated Ed. $1 60

Young's Night Thoughts . . . . do. . . 1 60

Cowper's Task, Table Talk, &c. . do. . . 1 60

Thomson's Seasons . . . . . . do. . . 1 60

Pollok's Course of Time . . . . do. . . 1 60

> These great moral poems are known wherever the English language is read, and are regarded as models of the best and purest literature. The books are beautifully illustrated, and notes explain all doubtful meanings, and furnish other matter of interest to the general reader.

Lord Bacon's Essays, (Boyd's Edition.) . . . 1 60

> Another grand English classic, affording the highest example of purity in language and style.

The Iliad of Homer.   Translated by POPE. . . .80

> Those who are unable to read this greatest of ancient writers in the original, should not fail to avail themselves of this metrical version by an eminent scholar and poet.

The Poets of Connecticut—Everest 1 75

> With the biographical sketches, this volume forms a complete history of the poetical literature of the State.

The Son of a Genius—Hofland 75

> A juvenile classic which never wears out, and finds many interested readers in every generation of youth.

Lady Willoughby 1 00

> The diary of a wife and mother, An historical romance of the seventeenth century. At once beautiful and pathetic, entertaining and instructive.

The Rhyming Dictionary—Walker 1 25

> A serviceable manual to composers of rhythmical matter, being a complete index of allowable rhymes.

LIBRARY OF REFERENCE.

Home Cyclopædia of Chronology $2 25
An index to the sources of knowledge—a dictionary of dates.

Home Cyclopædia of Geography 2 25
A complete gazetteer of the world.

Home Cyclopædia of Useful Arts 2 25
Covering the principles and practice of modern scientific enterprise, with a record of important inventions in agriculture, architecture, domestic economy, engineering, machinery, manufactures, mining, photogenic and telegraphic art, &c., &c.

Home Cyclopædia of Literature & Fine Arts 2 25
A complete index to all terms employed in belles lettres, philosophy, theology, law, mythology, painting, music, sculpture, architecture, and all kindred arts.

LIBRARY OF TRAVEL.

Ship and Shore—Colton 1 50
In Madeira, Lisbon, and the Mediterranean Ocean. Illustrated.

Land and Lee—Colton 1 50
In the Bosphorus and Ægean. Illustrated.

Sea and Sailor—Colton 1 50
Notes on France and Italy. Illustrated.

Deck and Port—Colton 1 50
A cruise to California. Illustrated.

Three Years in California—Colton 1 50
During the gold fever. Illustrated.
These racy descriptions of travel are regarded as models in this department of literature. They are read by old and young with vast interest and profit.

A Visit to Europe—Silliman, 2 vols. 3 00
A very spicy book of foreign travel. It brings every opportunity of the tourist to the feet of the reader.

TRAVEL—Continued.

Life in the Sandwich Islands—Cheever . .$1 50

The "heart of the Pacific, as it was and is," shows most vividly the contrast between the depth of degradation and barbarism, and the light and liberty of civilization, so rapidly realized in these islands under the humanizing influence of the Christian religion. Illustrated.

Peruvian Antiquities—Von Tschudi. . . . 1 50
Travels in Peru—Von Tschudi 1 50

The first of these volumes affords whatever information has been attained by travelers and men of science concerning the extinct people who once inhabited Peru, and who have left behind them many relics of a wonderful civilization. The "Travels" furnish valuable information concerning the country and its inhabitants as they now are. Illustrated.

Ancient Monasteries of the East—Curzon . 1 50

The exploration of these ancient seats of learning has thrown much light upon the researches of the historian, the philologist, and the theologian, as well as the general student of antiquity. Illustrated.

Discoveries in Babylon & Nineveh—Layard 1 75

Valuable alike for the information imparted with regard to these most interesting ruins, and the pleasant adventures and observations of the author in regions that to most men seem like Fairyland. Illustrated.

Egypt and the Holy Land—Spencer . . . 1 75

Still another volume of eastern travel. The many incontrovertible proofs of Scripture observed by the pains-taking modern traveler are worth the price of the book. Illustrated.

St. Petersburgh—Jermann 1 00

Americans are less familiar with the history and social customs of the Russian people than those of any other modern civilized nation. Opportunities such as this book affords are not, therefore, to be neglected.

The Polar Regions—Osborn 1 25

A thrilling and intensely interesting narrative of one of the famous expeditions in search of Sir John Franklin—unsuccessful in its main object, but adding many facts to the repertoire of science.

Thirteen Months in the Confederate Army 75

The author, a northern man conscripted into the Confederate service, and rising from the ranks by soldierly conduct to positions of responsibility, had remarkable opportunities for the acquisition of facts respecting the conduct of the Southern armies, and the policy and deeds of their leaders. He participated in many engagements, and his book is one of the most exciting narratives of adventure ever published. Mr. Stevenson takes no ground as a partizan, but views the whole subject as with the eye of a neutral—only interested in subserving the ends of history by the contribution of impartial facts. Illustrated.

LIBRARY OF HISTORY.

History of Europe—Alison $2 50

A reliable and standard work, which covers with clear, connected, and complete narrative, the eventful occurrences transpiring from A. D. 1789 to 1815, being mainly a history of the career of Napoleon Bonaparte.

History of England—Berard 1 75

Combining a history of the social life of the English people with that of the civil and military transactions of the realm.

History of Rome—Ricord 1 60

Possesses all the charm of an attractive romance. The fables with which this history abounds are introduced in such away as not to deceive the inexperienced reader, while adding vastly to the interest of the work and affording a pleasing index to the genius of the Roman people. Illustrated.

The Republic of America—Willard 2 25

Universal History in Perspective—Willard 2 25

From these two comparatively brief treatises the intelligent mind may obtain a comprehensive knowledge of the history of the world in both hemispheres. Mrs. Willard's reputation as an historian is wide as the land. Illustrated.

Ecclesiastical History—Marsh 2 00

A history of the Church in all ages, with a comprehensive review of all forms of religion from the creation of the world. No other source affords, in the same compass, the information here conveyed.

History of the Ancient Hebrews—Mills . . 1 75

The record of "God's people" from the call of Abraham to the destruction of Jerusalem; gathered from sources sacred and profane.

The Mexican War—Mansfield 1 50

A history of its origin, and a detailed account of its victories; with official dispatches, the treaty of peace, and valuable tables. Illustrated.

Early History of Michigan—Sheldon . . . 1 75

A work of value and deep interest to the people of the West. Compiled under the supervision of Hon. Lewis Cass. Embellished with portraits.

LIBRARY OF BIOGRAPHY.

Life of Dr. Sam. Johnson—Boswell . . .$2 25

This work has been before the public for seventy years, with increasing approbation. Boswell is known as " the prince of biographers."

Henry Clay's Life and Speeches—Mallory
2 vols. 4 50

This great American statesman commands the admiration, and his character and deeds solicit the study of every patriot.

Life & Services of General Scott—Mansfield 1 75

The hero of the Mexican war, who was for many years the most prominent figure in American military circles, should not be forgotten in the whirl of more recent events than those by which he signalized himself. Illustrated.

Garibaldi's Autobiography 1 50

The Italian patriot's record of his own life, translated and edited by his friend and admirer. A thrilling narrative of a romantic career. With portrait.

Lives of the Signers—Dwight 1 50

The memory of the noble men who declared our country free at the peril of their own "lives, fortunes, and sacred honor," should be embalmed in every American's heart.

Life of Sir Joshua Reynolds—Cunningham 1 50

A candid, truthful, and appreciative memoir of the great painter, with a compilation of his discourses. The volume is a text-book for artists, as well as those who would acquire the rudiments of art. With a portrait.

Prison Life 75

Interesting biographies of celebrated prisoners and martyrs, designed especially for the instruction and cultivation of youth.

LIBRARY OF NATURAL SCIENCE.

The Treasury of Knowledge $1 25

A cyclopædia of ten thousand common things, embracing the widest range of subject-matter. Illustrated.

Ganot's Popular Physics 1 75

The elements of natural philosophy for both student and the general reader. The original work is celebrated for the magnificent character of its illustrations, all of which are literally reproduced here.

Principles of Chemistry—Porter 2 00

A work which commends itself to the amateur in science by its extreme simplicity, and careful avoidance of unnecessary detail. Illustrated.

Class-Book of Botany—Wood 3 50

Indispensable as a work of reference. Illustrated.

The Laws of Health—Jarvis 1 65

This is not an abstract *anatomy*, but all its teachings are directed to the best methods of preserving health, as inculcated by an intelligent knowledge of the structure and needs of the human body. Illustrated.

Vegetable & Animal Physiology—Hamilton 1 25

An exhaustive analysis of the conditions of life in all animate nature. Illustrated.

Elements of Zoology—Chambers 1 50

A complete view of the animal kingdom as a portion of external nature. Illustrated.

Astronography—Willard 1 00

The elements of astronomy in a compact and readable form. Illustrated.

Elements of Geology—Page 1 25

The subject presented in its two aspects of interesting and important. Illustrated.

Lectures on Natural History—Chadbourne 75

The subject is here considered in its relations to intellect, taste, health, and religion.

VALUABLE LIBRARY BOOKS.

The Political Manual—Mansfield $1 25

Every American youth should be familiar with the principles of the government under which he lives, especially as the policy of this country will one day call upon him to participate in it, at least to the extent of his ballot.

American Institutions—De Tocqueville . . 1 50

Democracy in America—De Tocqueville . . 2 25

The views of this distinguished foreigner on the genius of our political institutions are of unquestionable value, as proceeding from a standpoint whence we seldom have an opportunity to hear.

Constitutions of the United States 2 52

Contains the Constitution of the General Government, and of the several State Governments, the Declaration of Independence, and other important documents relating to American history. Indispensable as a work of reference.

Public Economy of the United States . . . 2 25

A full discussion of the relations of the United States with other nations, especially the feasibility of a free-trade policy.

Grecian and Roman·Mythology—Dwight . 2 25

The presentation, in a systematic form, of the Fables of Antiquity, affords most entertaining reading, and is valuable to all as an index to the mythological allusions so frequent in literature, as well as to students of the classics who would peruse intelligently the classical authors. Illustrated.

Modern Philology—Dwight 1 75

The science of language is here placed, in the limits of a moderate volume, within the reach of all.

General View of the Fine Arts—Huntington 1 75

The preparation of this work was suggested by the interested inquiries of a group of young people, concerning the productions and styles of the great masters of art, whose names only were familiar. This statement is sufficient index of its character.

Morals for the Young—Willard . 75

A series of moral stories, by one of the most experienced of American educators. Illustrated.

Improvement of the Mind—Isaac Watts · · 45

A classical standard. No young person should grow up without having perused it.

A. S. Barnes & Company

[From the NEW YORK PATHFINDER, Aug. 1866.]

This well-known and long-established Book and Stationery House has recently removed from the premises with which it has been identified for over twenty years, to the fine buildings, Nos. 111 and 113 William Street, corner of John Street, New York, one block only from the old store. Here they have been enabled to organize their extensive business in all its departments more thoroughly than ever before, and enjoy. facilities possessed by no other house in New York, for handling in large quantities and at satisfactory prices every thing in their line.

A visit to this large establishment will well repay the curious. On entering, we find the first floor occupied mainly by offices appertaining to the different departments of the business. The first encountered is the "Salesman's Office," where attentive young men are always in waiting to supply the wants of customers. Further on we come to the Entry Department, where all invoices from the several sales-rooms are collected and recorded. Next comes the General Office of the firm. Then a modest sign indicates the entrance to the "Teachers' Reading-Room"—a spacious and inviting apartment set apart for the use of the many professional friends and visitors of this house. On the table we noticed files of educational journals and other periodical matter—while a book-case contains a fine selection of popular publications as samples. The private office of the senior partner, and the Book-keeper's and Mailing Clerk's respective apartments, are next in order, and complete the list of offices on this floor. The remainder of the space is occupied by the departments of stock known as "Late Publications" and "General School Books."

Descending to the finely lighted and ventilated basement, we find the "Exchange Trade," "Shipping," and "Packing" departments. Here, also, is kept a heavy stock of the publications of the house, while a series of vaults under the sidewalk afford accommodation for a variety of heavy goods. Stepping on the platform of the fine Otis' Steam-elevator, which runs from bottom to top of the building, the visitor ascends to the

Second Story.—This floor is occupied by the Blank Book and Stationery Department, where are carried on all the details of an entirely separate business, by clerks especially trained in this line. Here every thing in the way of imported and domestic stationery is kept in vast assortment and to suit the wants of every class of trade. The system of organization mentioned above enables this house to compete successfully with those who make this branch a specialty, while the convenience to Booksellers of making all their purchases at one place is indisputable.

On the third floor are found the following varieties of stock: Toy and Juvenile Books, Bibles and Prayer Books, Standard Works, Photograph Albums, &c. The fourth and fifth stories are occupied as store-rooms for Standard School Stock. During the summer, while all the manufacturing energies of the concern are devoted to the preparation and accumulation of stock for the fall trade, upwards of *half a million of volumes* are gathered in these capacious rooms at once.

The manufacturing department of this house is carried on in the old premises, Nos. 51, 53, and 55 John Street, and 2, 4, and 6 Dutch Street. A large number of operatives, with adequate presses and machinery, are constantly employed in turning out the popular publications of the firm.

The Peabody Correspondence.

NEW YORK, *April 29, 1867.*

TO THE BOARD OF TRUSTEES OF THE PEABODY EDUCATIONAL FUND:

GENTLEMEN—Having been for many years intimately connected with the educational interests of the South, we are desirous of expressing our appreciation of the noble charity which you represent. The Peabody Fund, to encourage and aid common schools in these war-desolated States, can not fail of accomplishing a great and good work, the beneficent results of which, as they will be exhibited in the future, not only of the stricken population of the South, but of the nation at large, seem almost incalculable.

It is probable that the use of meritorious text-books will prove a most effective agency toward the thorough accomplishment of Mr. Peabody's benevolent design. As we publish many which are considered such, we have selected from our list some of the most valuable, and ask the privilege of placing them in your hands for gratuitous distribution in connection with the fund of which you have charge, among the teachers and in the schools of the destitute South.

Observing that the training of teachers (through the agency of Normal Schools and otherwise) is to be a prominent feature of your undertaking, we offer you for this purpose 5,000 volumes of the "Teachers' Library,"—a series of professional works designed for the efficient self-education of those who are in their turn to teach others—as follows:—

500 Page's Theory and Practice of Teaching.	250 Bates' Method of Teachers' Institutes.
500 Welch's Manual of Object-Lessons.	250 De Tocqueville's American Institut'ns.
500 Davies' Outlines of Mathematical Science.	250 Dwight's Higher Christian Education.
	250 History of Education.
250 Holbrook's Normal Methods of Teaching.	250 Mansfield on American Education.
	250 Mayhew on Universal Education.
250 Wells on Graded Schools.	250 Northend's Teachers' Assistant.
250 Jewell on School Government.	250 Northend's Teacher and Parent.
250 Fowle's Teachers' Institute.	250 Root on School Amusements.
	250 Stone's Teachers' Examiner.

In addition to these we also ask that you will accept 25,000 volumes of school-books for intermediate classes, embracing—

5,000 The National Second Reader.	5,000 Beers' Penmanship.
5,000 Davies' Written Arithmetic.	500 First Book of Science.
5,000 Monteith's Second Book in Geography.	500 Jarvis' Physiology and Health.
	500 Peck's Ganot's Natural Philosophy.
3,000 Monteith's United States History.	500 Smith & Martin's Book-keeping.

Should your Board consent to undertake the distribution of these volumes, we shall hold ourselves in readiness to pack and ship the same in such quantities and to such points as you may designate.

We further propose that, should you find it advisable to use a greater quantity of our publications in the prosecution of your plans, we will donate, for the benefit of this cause, *twenty-five per cent.* of the usual wholesale price of the books needed.

Hoping that our request will meet with your approval, and that we may have the pleasure of contributing in this way to wants with which we deeply sympathize, we are, gentlemen, very respectfully yours, A. S. BARNES & CO.

BOSTON, *May 7, 1867.*

MESSRS. A. S. BARNES & CO., PUBLISHERS, NEW YORK:

GENTLEMEN—Your communication of the 29th ult., addressed to the Trustees of the Peabody Education Fund, has been handed to me by our general agent, the Rev. Dr. Sears. I shall take the greatest pleasure in laying it before the board at their earliest meeting. I am unwilling, however, to postpone its acknowledgment so long, and hasten to assure you of the high value which I place upon your gift. Five thousand volumes of your "Teachers' Library," and twenty-five thousand volumes of "School-Books for intermediate classes," make up a most munificent contribution to the cause of Southern education in which we are engaged. Dr. Sears is well acquainted with the books you have so generously offered us, and unites with me in the highest appreciation of the gift. You will be glad to know, too, that your letter reached us in season to be communicated to Mr. Peabody, before he embarked for England on the 1st instant, and that he expressed the greatest gratification and gratitude on hearing what you had offered.

Believe me, gentlemen, with the highest respect and regard, your obliged and obedient servant, ROBT. C. WINTHROP, Chairman.

www.ingramcontent.com/pod-product-compliance
Lightning Source LLC
Chambersburg PA
CBHW031420020726
47499CB00005B/1514